ADVANCES IN CEREBRAL SPECT IMAGING

AN ATLAS AND GUIDELINE FOR PRACTITIONERS

EDITORS:

Ronald L. Van Heertum, M.D.
Professor of Clinical Radiology
NY Medical College, Valhalla, NY
Chief of Nuclear Medicine
St. Vincent's Hospital and
Medical Center
New York, NY

Ronald S. Tikofsky, Ph.D.
Research Associate Professor
of Radiology
Division of Nuclear Medicine
Medical College of Wisconsin
Milwaukee, WI

CONSULTING EDITORS:

David L. Daniels, M.D.
Associate Professor of Radiology
Medical College of Wisconsin
Milwaukee, WI

Charles R. Noback, Ph.D.
Professor Emeritus
Department of Anatomy
College of Physicians and Surgeons
Columbia University
New York, NY

R. Anthony O'Connell, M.D.
Professor of Clinical Psychiatry
NY Medical College, Valhalla, NY
Clinical Director
Department of Psychiatry
St. Vincent's Hospital and
Medical Center
New York, NY

Alan B. Rubens, M.D.
Professor and Head
Department of Neurology
College of Medicine, University
of Arizona
Tucson, AZ

PUBLISHED BY:

TRIVIRUM PUBLISHING COMPANY
211 E.43rd St., New York, NY 10017 (212)986-0183

Distributed worldwide by Lea & Febiger • Philadelphia • London

Certain materials from cases in this atlas have appeared in print and are used here with permission from the publishers.

Dementias:
Case 1/Page 65: The Journal of Neuropsychiatry and Clinical Neurosciences; May 1989. American Psychiatric Press, Inc., Washington, D.C.
Case 2/Page 66: Directions in Psychiatry; Volume 8, 1988. Hatherleigh Co. Ltd., New York.
Case 12/Page 76: Advances in Funcional Neuroimaging; Volume 1, No. 2, 1988. Macmillan Healthcare Information, New Jersey.

Seizure Disorders:
Case 2/Page 81: Advances in Functional Brain Imaging; Volume 1, No. 3, 1988. Macmillan Healthcare Information, New Jersey.
Case 3/Page 82: Applied Radiology; June 1988. Romaine Pierson Publishing Co., New York.

Psychiatry:
Case 1/Page 90: The Journal of Neuropsychiatry and Clinical Neurosciences; May 1989. American Psychiatric Press, Inc. Washington, D.C.
Case 3/Page 92-93: Advances in Functional Neuroimaging; Volume 1, No. 1. Macmillan Healthcare Information, New Jersey.
Case 4/Page 94: Clinical Nuclear Medicine; May 1989. J.B. Lippincott Co., Pennsylvania
Case 5/Page 95: The Journal of Neuropsychiatry and Clinical Neurosciences; May 1989. American Psychiatric Press, Inc. Washington, D.C.
Case 6/Page 96: Advances in Functional Neuroimaging; Volume 1, No. 2. Macmillan Healthcare Information, New Jersey.

Miscellaneous:
Case 1/Page 98: Advances in Functional Neuroimaging; Volume 1, No. 2. Macmillan Healthcare Information, New Jersey.

The co-medical editors and medical editorial consultants have had complete independence in the selection of materials and cases presented. The opinions or recommendations expressed are those of such editors and consultants and are not to be construed as the opinions or recommendations of Medi-Physics, Inc. or Trivirum Publishing Company. Practitioners are urged to consult the full prescribing information on any drug or device mentioned.

ISBN 0-8121-1260-1

Designed/Produced by Medical Multimedia Corp. and published by Trivirum Publishing Co. under an educational grant provided by Medi-Physics, Inc.

Table of Contents

Introduction

This atlas is designed to introduce functional brain imaging to imaging clinicians (in nuclear medicine and radiology) and to nonimaging clinicians (neurologists, neurosurgeons, and psychiatrists), using Single Photon Emission Computed Tomography (SPECT) and newer extractable lipophilic radiopharmaceuticals such as I^{123}-labelled n-isopropyl iodoamphetamine (IMP), and to show how to perform and interpret the studies. When this atlas was in production, IMP was the only FDA-approved agent available in the United States; for this reason, only those studies done with IMP were included here. As cerebral SPECT evolves, future editions will expand on this.

To provide the greatest utility, the text has been organized into three sections, with clinical indications and case examples. The general aspects of the imaging technique are covered in Section I, including patient preparation, acquisition, processing, and display; normal anatomy is discussed, illustrated, and correlated with the anatomy visualized on SPECT and MRI studies, thereby demonstrating the value and ease of cross correlation between high-resolution structural studies and regional functional imaging. Helpful hints for interpreting the functional studies are included. The last part of this section examines the pitfalls imposed by the patient and by technical and interpretive factors.

Section II covers disease states, with the emphasis on cerebrovascular diseases, the dementias, seizure disorders, psychiatric disorders, trauma, and brain tumors. Multiple examples—with CT and MRI examinations—have been included to illustrate the relative strengths and weaknesses of this new imaging modality.

Section III includes an Appendix describing the specifics of the procedures for cerebral SPECT imaging with the various types of equipment now available. A discussion of quality control of radiopharmaceuticals and equipment is also found in this Section, as is a Bibliography and an Index.

The editors hope this atlas will be useful and interesting, and that its information can be readily incorporated into your practice.

Acknowledgement

We would like to extend our appreciation to all those physicians, radiologists, and technicians who, in using SPECT, have advanced it to its present functional state; their continuing efforts will insure that the technique will impose few limitations.

We would like to thank Stanley Waine and the staff at Medical Multimedia Corporation, and the people at Trivirum Publishing Company, for their efforts in producing this work.

We thank our wives, Jacqueline and Rita, and our children for their patience and forbearance. To them we dedicate this work.

Ronald L. Van Heertum and
Ronald S. Tikofsky, Editors

Contributing Authors

Amjad Ali, M.D.	*Associate Professor of Diagnostic Radiology and Nuclear Medicine, Rush Medical College and Rush-Presbyterian/St. Luke's Medical Center, Chicago, IL*
William L. Ashburn, M.D.	*Professor of Radiology and Director, Division of Nuclear Medicine, UCSD Medical Center, San Diego, CA*
Marvin B. Cohen, M.D.	*Professor of Medicine and Radiology, UCLA Medical Center, Chief of Nuclear Medicine Service, VA Medical Center, Sepulveda, CA*
B. David Collier, M.D.	*Associate Professor of Radiology, Director of Nuclear Medicine, Medical College of Wisconsin, Milwaukee, WI*
Donald Fox, M.D.	*Radiologist, San Jose Imaging Center, San Jose, CA*
Harvey L. Hecht, M.D.	*Clinical Assistant Professor of Radiology, Columbia-Presbyterian Medical Center, Radiologist, Stamford Hospital, Stamford, CT*
Robert S. Hellman, M.D.	*Assistant Professor of Radiology, Medical College of Wisconsin, Milwaukee, WI*
Thomas C. Hill, M.D.	*Associate Professor of Radiology, Harvard Medical School, Director of Nuclear Medicine, New England Deaconess Hospital, Boston, MA*
Robert L. Kagan, M.D.	*Director, MRI Scan Center, Northeast Medical Center, Fort Lauderdale, FL*

Lee Katims, M.D. *Chief of Radiology, West Boca Medical Center, Boca Raton, FL*

David L. Kaufman, M.D. *Attending Physician, St. Vincent's Hospital and Medical Center, New York, N.Y.*

Kazufumi Kimura, M.D. *Associate Professor, Department of Nuclear Medicine, Biomedical Research Center, Osaka University Medical School, Osaka, Japan*

Leighton P. Mark, M.D. *Assistant Professor of Radiology, Medical College of Wisconsin, Milwaukee, WI*

Alan H. Maurer, M.D. *Associate Professor of Diagnostic Imaging, Director of Nuclear Medicine, Temple University, Philadelphia, PA*

Caesar Mayo, M.D. *Neurologist and President, San Jose Imaging Center, San Jose, CA*

Joji Nagawara, M.D. *Neurosurgeon, Department of Neurosurgery, Nakamura Memorial Hospital, Sopporo, Japan*

SECTION I: A

Technique

Patient Preparation

Before or when the patient arrives for the study, some of the interview time should be devoted to explaining the procedure and allaying the patient's concerns about it. It is also advisable to use this time to obtain a brief history, including information concerning handedness. This type of information is particularly useful when evaluating patients with cerebrovascular disease (CVD).

In some instances, particularly when dealing with dementia patients, this sort of communication will be best accomplished with other members of the patient's family. Although spending this advance time with the patient is important with any type of nuclear medicine study, this approach is absolutely essential to ensure a successful cerebral SPECT study.

Following the orientation, the patient should be given three drops of a saturated potassium iodide solution in orange juice or similar beverage. This preparation is necessary to minimize thyroid gland irradiation from uptake of the radioactive I^{123}. Prior to the injection, the patient should be requested to void.[1]

After completing the orientation and preparation procedures, the patient is comfortably positioned on the imaging couch with his arms resting at his sides. In some systems, the imaging couches have removable wing armrests that can be used to allow greater ease of positioning and patient comfort during the injection of the radiopharmaceutical. The imaging room environment should be carefully controlled with low noise, low light, and minimal traffic. The patient should be instructed, whenever possible, to keep his eyes open during the injection and the ensuing equilibration period (15 to 20 minutes post injection). A dose of 3 to 5 (111 to 185 mBq) mCi I^{123}-labeled n-isopropyl iodoamphetamine is administered intravenously. Following the radiopharmaceutical injection, the patient should remain in a quiet, comfortable, motionless position for the next 20 minutes. During this time, the technologist begins the actual setup and positioning of the detector system so that the image acquisition can begin immediately thereafter.

Acquisition Technique

Specific acquisition techniques will vary depending on the imaging equipment used. There are, however, a number of considerations that apply to cerebral SPECT studies no matter who the manufacturer. In all instances, the patient should be comfortably positioned on the imaging couch with his head "immobilized" in a radiolucent head-holder. The patient's head is fixed so that a line perpendicular to the detector head runs $5°$ to $7°$ cephalad to the canthomeatal line (Fig. 1). Furthermore, the patient and detector should be positioned in such a manner that the imaging device remains parallel to the coronal plane at the 0 ° and 180 ° detector positions and parallel to the sagittal orthogonal plane at the 90 ° and 270 ° positions (Fig. 2). It is essential also that the patient be centered so that his brain remains within the field of view on all of the individually acquired projection images. In addition, the detector should be positioned as close to the patient's brain as possible, preferably with a radius of rotation of 14 cm or less from the surface of the collimator to the center of the patient's brain. This tight radius of rotation can be achieved by using a variety of approaches, including a cut-off detector head, long-bore fan beam collimation, and frontal tomography[2,3,4] (Fig. 3A,B,C). The cut-off or "shaved" detector is specifically designed so that the detector will clear the patient's shoulders. When a cut-off detector is not available, alternative approaches include either long-bore fan beam collimation or frontal tomography —both of which have been shown to be useful for cerebral SPECT imaging studies.

Following proper positioning of the patient and detector, the actual acquisition procedure can begin. In general, 64 to 128 projection images are acquired, for a total imaging-time of 30 to 45 minutes. It is recommended that acquisition begin with the detector positioned to the side of the patient's head, in a clockwise fashion (Fig. 4).

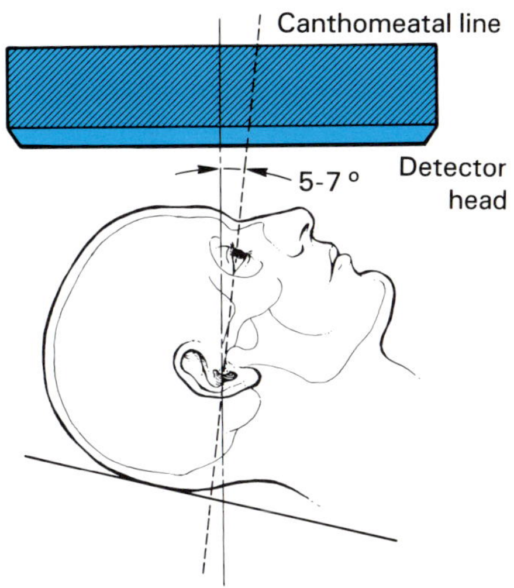

Figure 1. Detector head in relation to canthomeatal line.

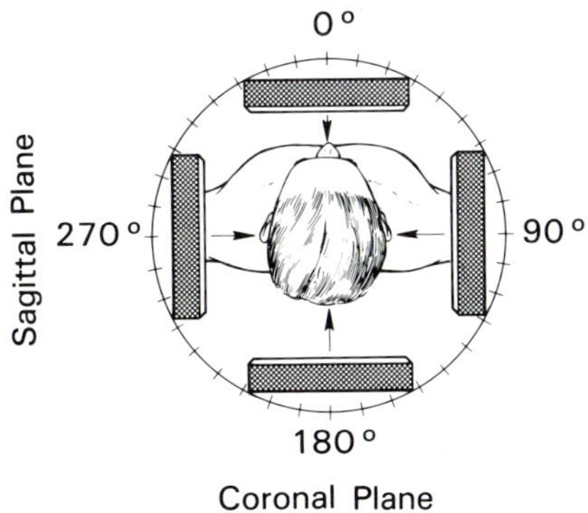

Figure 2. Relation of detector to patient's head at 0, 90, 180, and 270 degrees.

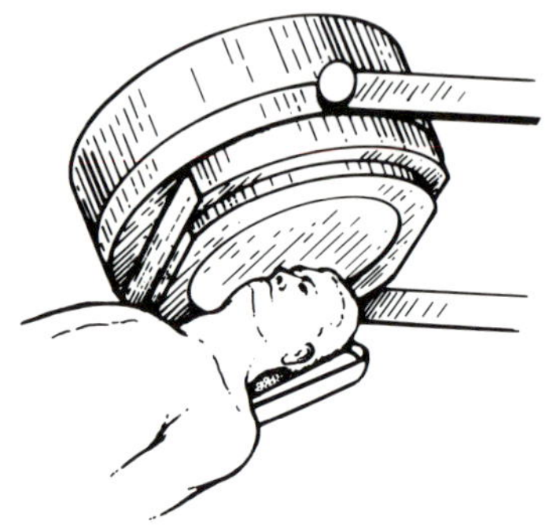

Figure 3. A: Cut-off head configuration.

Figure 3. B: Long-bore collimator.

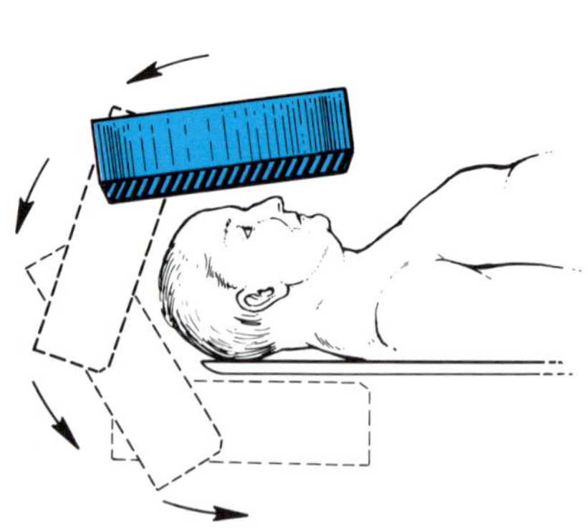

Figure 3. C: Frontal tomography

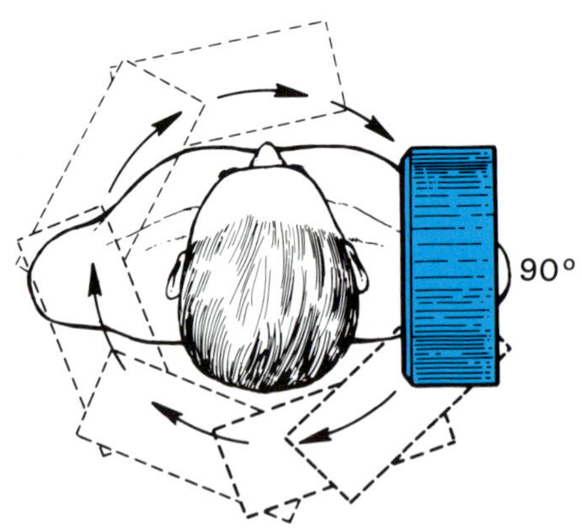

Figure 4. Clockwise rotation of detector head starting on patient's right side.

This image-acquisition technique is particularly useful with anxious, agitated, or claustrophobic patients, since the detector first rotates posteriorly and so allows the patient to become acclimated to the procedure before the detector moves anteriorly over his face at the end of the procedure. A 360° rotation orbit should be used whenever possible. In compromise situations—as with difficult or agitated patients—a reduced imaging-time for each projection image will usually prove to be a viable solution.

The type of collimation for the procedure will vary depending on the equipment used. To ensure that a proper study has been obtained, the projection images should be evaluated before the patient leaves the department. A useful method for performing this assessment is to display the acquired projection images in a closed-loop cine display. An alternative is to use a horizontal profile ROI analysis of the acquired data to generate a sinogram display[5]. (These approaches are discussed in detail in the Pitfalls section.)

Processing Technique

After the study has been acquired and assessed for adequacy, the preprocessing and reconstruction procedures are performed. The images should be routinely uniformity corrected and reconstructed with attenuation correction. Specific processing protocols will vary from system to system. (The specifics of individual manufacturers' protocols are discussed in detail in the Appendix.) In general, the images are either spatially filtered before (preprocessed) or after (postprocessed) the reconstruction process, using a filtered back-projection technique. The reconstructed transaxial sections are then reoriented and displayed in the coronal and sagittal planes.

Display Technique

All three orthogonal imaging planes should be displayed. The transaxial images are generally displayed in the same format as transmission CT or MRI scans. The sections begin inferiorly and continue superiorly in a contiguous manner (Fig. 5). The coronal and sagittal planes are displayed in the same manner as MRI studies (Figs. 6 and 7). (These presentations are reviewed in greater detail in the Anatomic Correlation section.)

Although the images are generally displayed and archived on transparency film, direct viewing of the images on the monitor screen can be very helpful, since additional information can be obtained from such observations. The three dimensional surface display is an alternative technique that may become more useful.

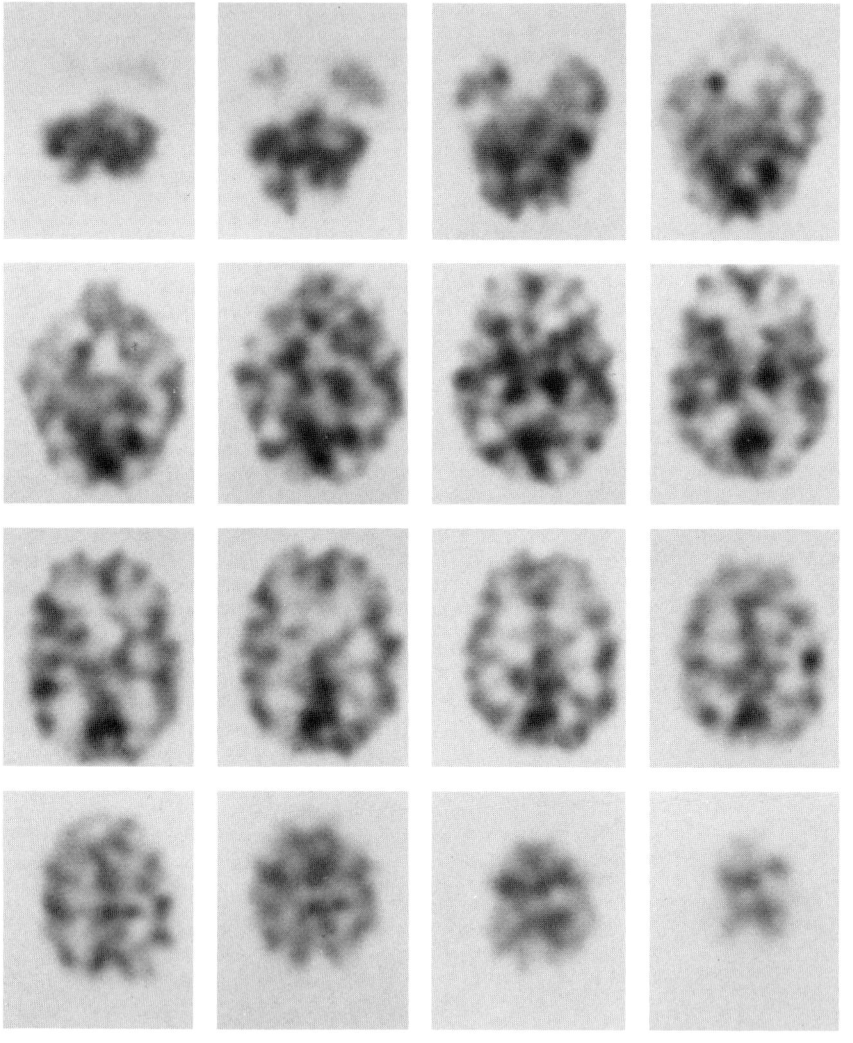

Figure 5. Normal cerebral SPECT images. Transaxial plane (inferior to superior).

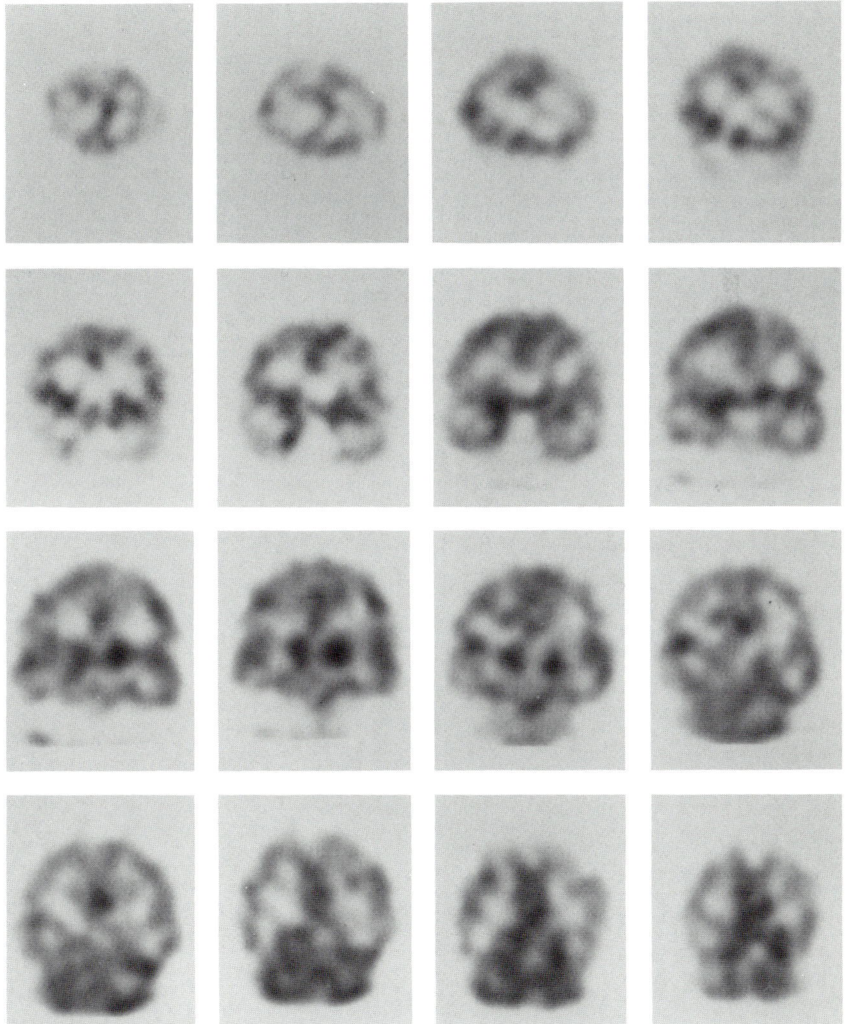

Figure 6. Normal cerebral SPECT images. Coronal plane (anterior to posterior).

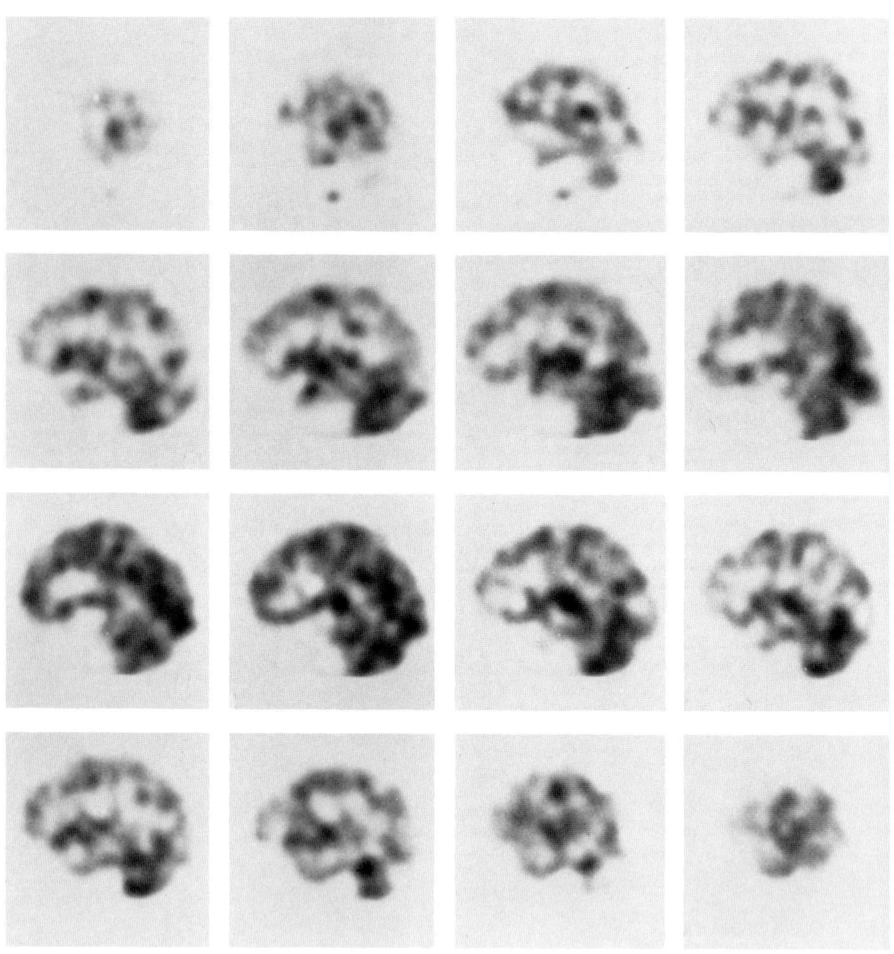

Figure 7. Normal cerebral SPECT images. Sagittal plane (right to left).

SECTION I: B

Normal & Correlative Anatomy

SECTION I: Normal/Correlative Anatomy

Oblique Lateral View of the Brain

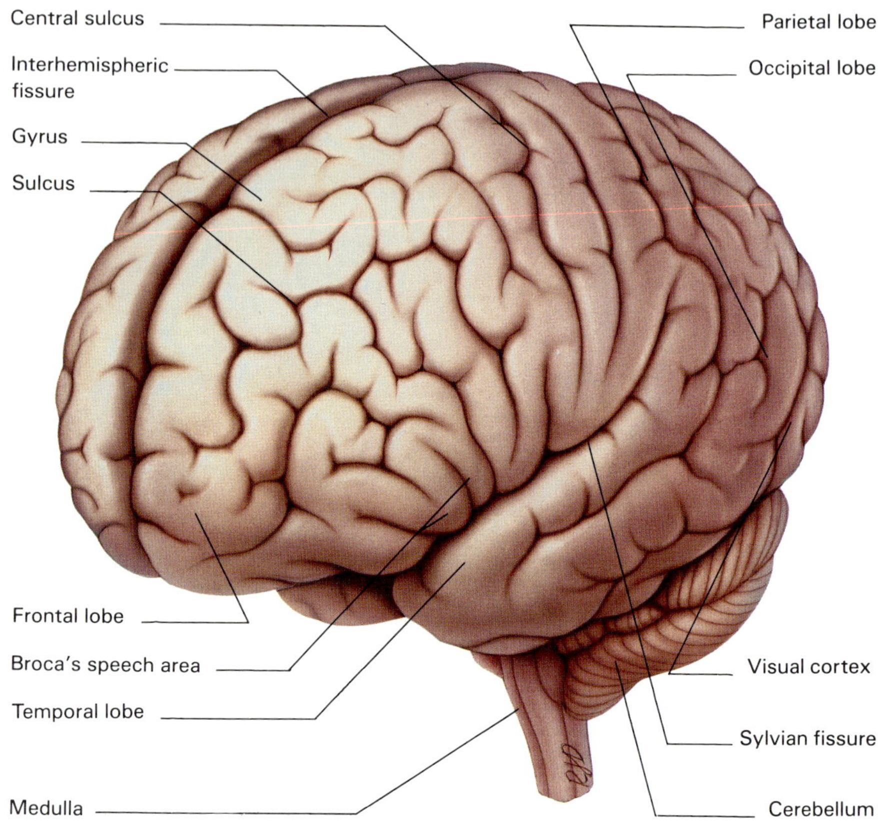

Surface Anatomy

The frontal, temporal, parietal, and occipital lobes are seen on the lateral view of the brain. The central sulcus (Fissure of Rolando) separates the frontal from parietal lobes. The Sylvian fissure divides the frontal from temporal and parietal lobes. There is no clear demarcation dividing the parietal from the occipital lobe. Functionally, the anterior portions of the frontal lobes are concerned with memory and emotional control, while the posterior portions are involved with the initiation and voluntary control of motor activity. The temporal lobe is involved with auditory function, and on the left side language comprehension; the parietal lobe is involved with the appreciation and interpretation of sensory input relating to form, shape, and weight. The occipital lobe is involved with primary visual function. The cerebellum is primarily involved with overall coordination of movement, including gait and balance. It is important to note that the two cerebral hemispheres do not have per-

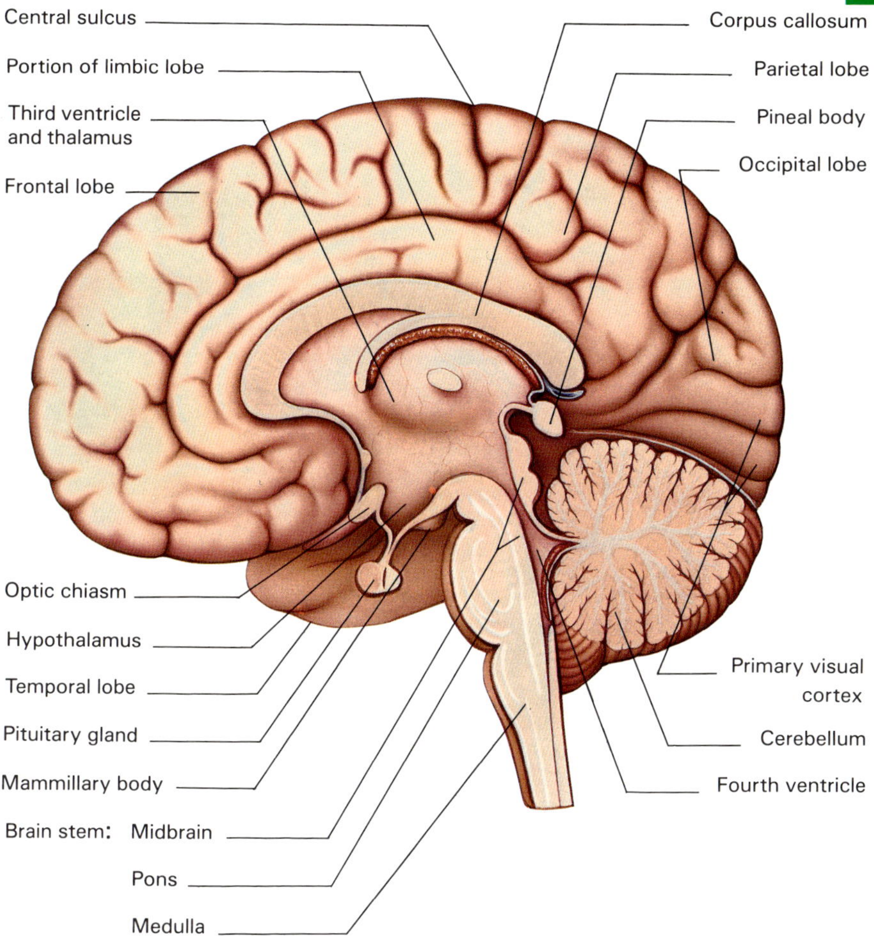

Central sulcus
Portion of limbic lobe
Third ventricle and thalamus
Frontal lobe
Corpus callosum
Parietal lobe
Pineal body
Occipital lobe
Optic chiasm
Hypothalamus
Temporal lobe
Pituitary gland
Mammillary body
Brain stem: Midbrain
Pons
Medulla
Primary visual cortex
Cerebellum
Fourth ventricle

Sagittal View of the Brain

fectly symmetric gross and microscopic anatomy, nor do they provide equivalent cognitive functions.

Sagittal View of the Brain

The two hemispheres are interconnected by a large band of nerve fibers, the corpus callosum. The limbic (fifth) lobe consisting of the cortex and associated structures superior to the corpus callosum is involved in emotions, drives, and behavioral expression.

The brain stem is involved with the control of respiration and cardiovascular activity. In addition the major sensorimotor pathways pass through the brain stem to and from the cortex. The cerebellum as noted above is involved with motor coordination, gait, and balance.

SECTION I: Normal/Correlative Anatomy

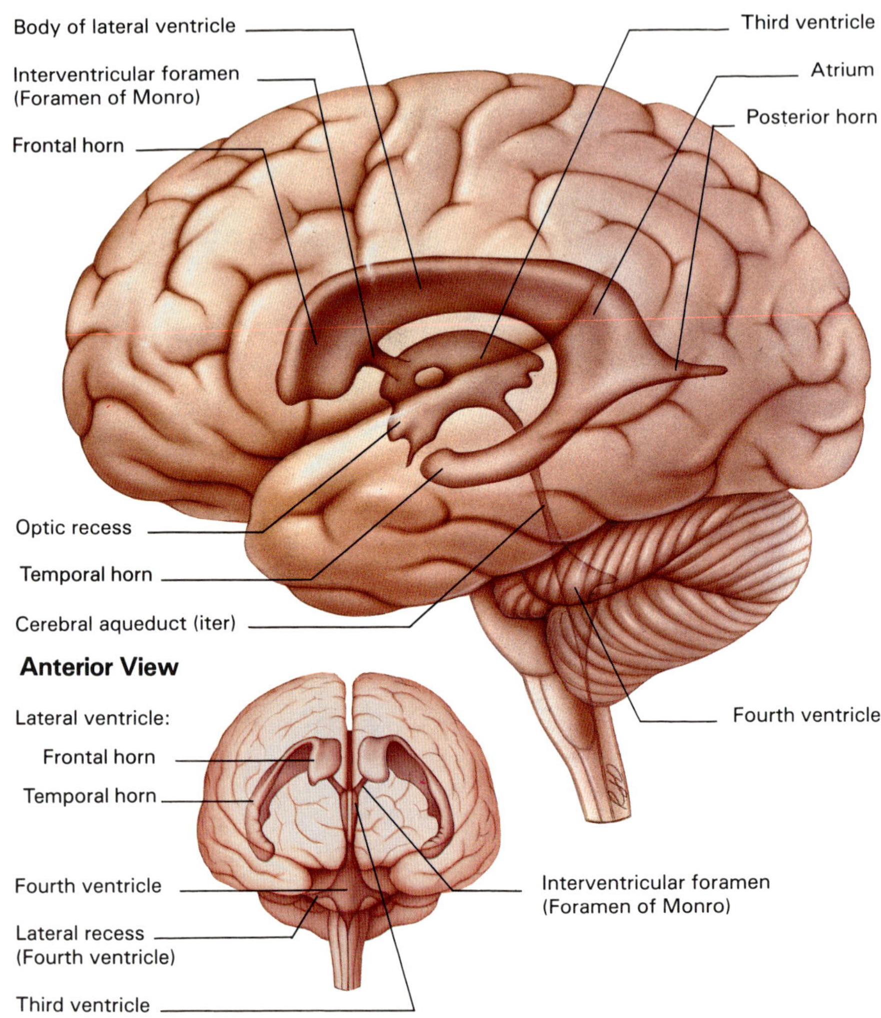

Ventricles: Lateral View

The Ventricular System

The ventricular system and subarachnoid space surrounding the brain, are filled with cerebral spinal fluid (CSF). The fluid is formed by the choroid plexus in each ventricle. CSF flows from the lateral ventricles through the interventricular foramen of Monro to the third ventricle, and then through the cerebral aqueduct and the fourth ventricle to the subarachnoid space. The system provides a fluid cushion upon which the brain "floats" protected from physical injury.

VASCULAR DISTRIBUTION: The Cerebral Arteries.

The arterial blood supply to the brain is derived from two major sources, i.e. the paired internal carotid (ICAs) and vertebral arteries. The vertebral arteries join to form the basilar artery, which in turn usually terminates by dividing into the posterior cerebral arteries (PCAs). Branches of the PCAs supply blood to the brain stem, cerebellum, the occipital lobes, and the inferior portion of the

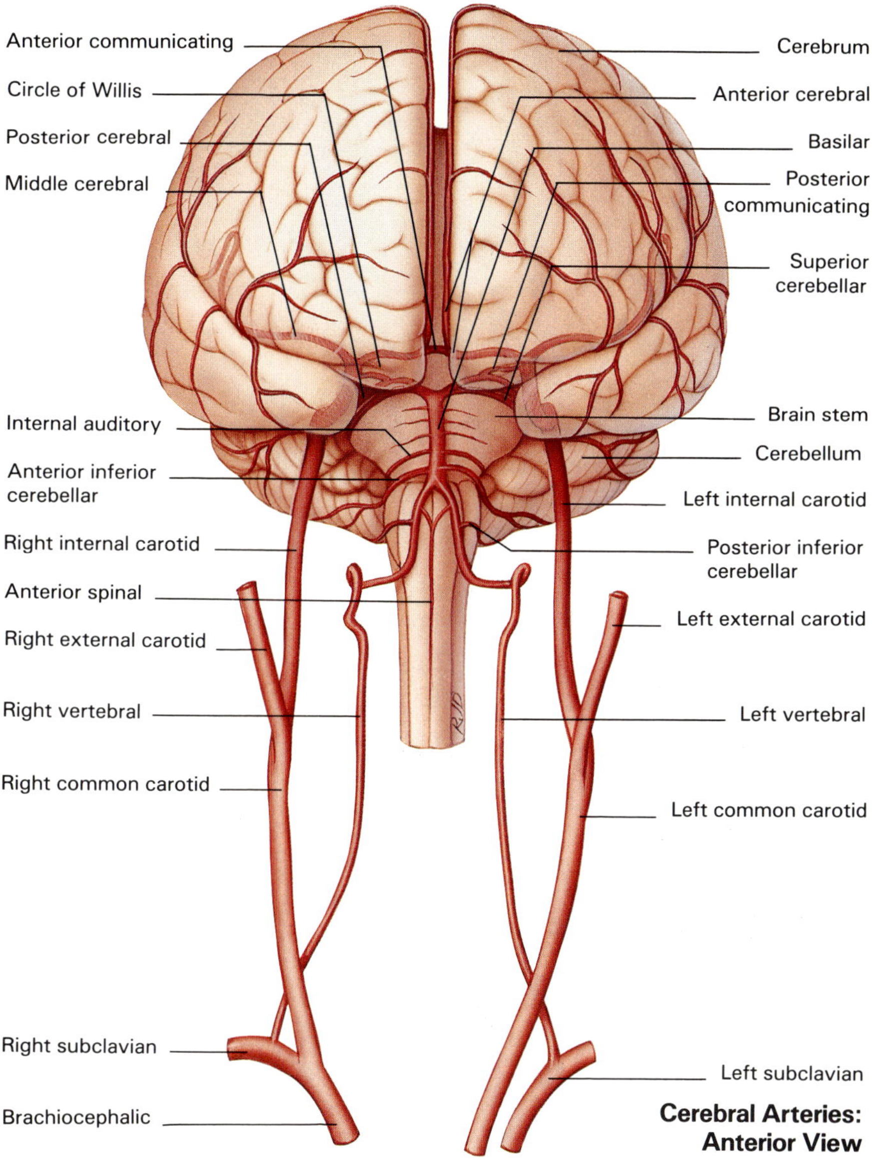

Cerebral Arteries: Anterior View

SECTION I: Normal/Correlative Anatomy

Frontal branches of middle cerebral artery
Branches of anterior cerebral artery
Frontal lobe
Parietal branch of middle cerebral artery
Parietal lobe
Occipital lobe
Branches of posterior cerebral artery
Temporal lobe
Temporal branches of middle cerebral artery
Sylvian fissure

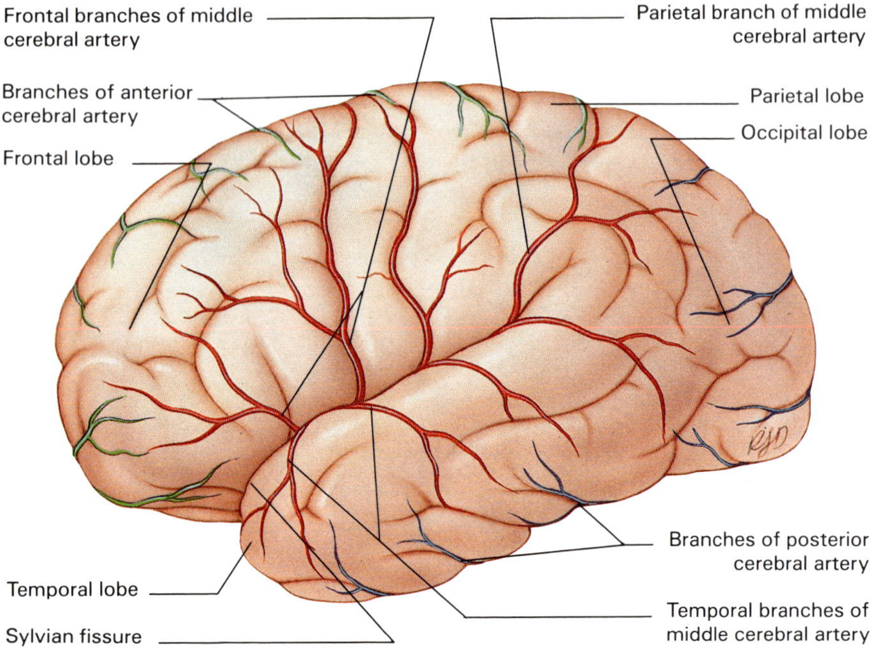

Lateral Surface of Left Hemisphere

Medial Surface of Right Hemisphere

Corpus callosum
Anterior cerebral artery
Frontal branches
Frontal lobe
Posterior cerebral artery
Calcarine branch
Parieto-occipital branch
Occipital lobe
Temporal lobe
Temporal branches of posterior cerebral artery
Temporal branches of middle cerebral artery
Internal carotid artery

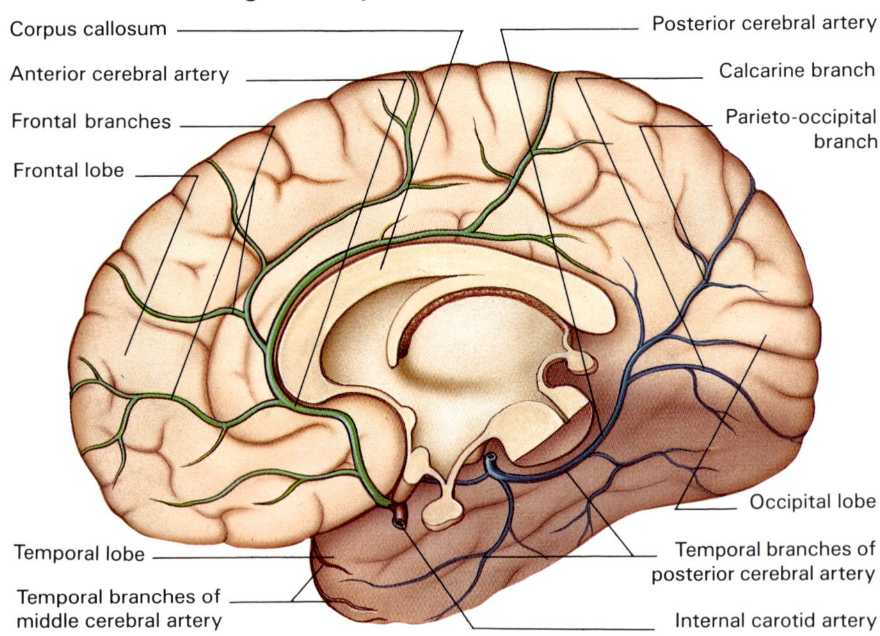

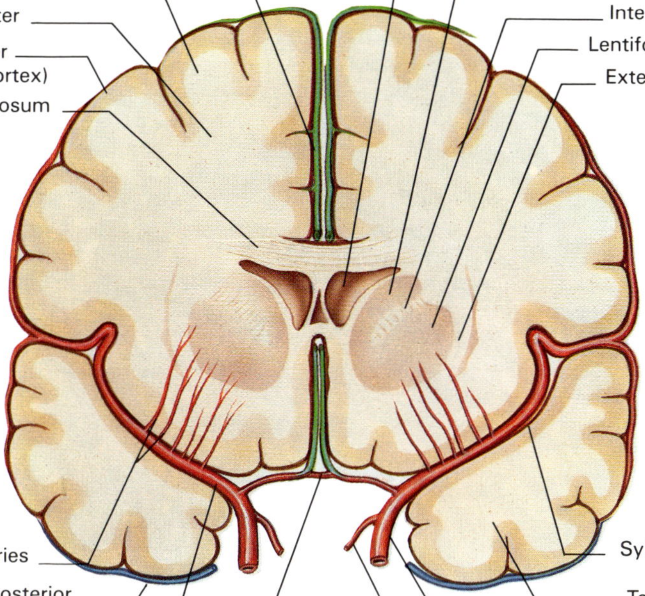

Anterior cerebral artery
Frontal lobe
White matter
Gray matter (cerebral cortex)
Corpus callosum
Lateral ventricle
Head of caudate nucleus
Internal capsule
Lentiform nucleus
External capsule
Striate arteries
Branch of posterior cerebral artery
Middle cerebral artery
Anterior communicating
Sylvian fissure
Temporal lobe
Internal carotid artery
Posterior communicating

Coronal Section Through the Cerebrum

temporal lobes. The ICAs usually terminate by dividing into anterior cerebral arteries (ACAs) and middle cerebral arteries (MCAs). The anterior and posterior circulations connect via the circle of Willis, an arterial ring found at the base of the brain.

As seen in the lateral view, the MCA extends laterally from the Sylvian fissure to supply regions of the frontal, temporal, parietal, and occipital lobes. However, the ACA provides the major arterial supply to the medial surface of the brain, in particular the superior surface of the corpus callosum with branches reaching to the frontal and parietal lobes. The PCA extends posteriorly and supplies the medial and posterior aspects of the temporal and occipital lobes.

On the coronal view, the MCAs are seen to extend between the temporal and parietal lobes. Their deep branches—striate arteries—supply the corpus striatum of the basal ganglia and the internal capsules.

SECTION I: Normal/Correlative Anatomy

CORRELATIVE ANATOMY

Cerebral SPECT studies are displayed in transaxial, coronal, and sagittal planes adjacent to the corresponding MR images.

The major structures of the cerebrum such as the frontal, temporal, parietal, and occipital lobes are identified. The cerebellar hemispheres and basal ganglia are also shown. However, the spatial resolution of SPECT images is less than that of MRI. Thus, SPECT images do not demonstrate the different structures of basal ganglia, brain stem, and the edge between white matter and the ventricles.

SPECT images should not be used to show detailed anatomy although they demonstrate the major anatomic landmarks. SPECT imaging—when correlated with anatomic studies such as MRI and CT—may be useful in delineating some pathological processes using multiple planes. More importantly, in some disease states, changes in blood flow and metabolism may only be appreciated with SPECT imaging, but not with CT or MRI.

Imaging Planes

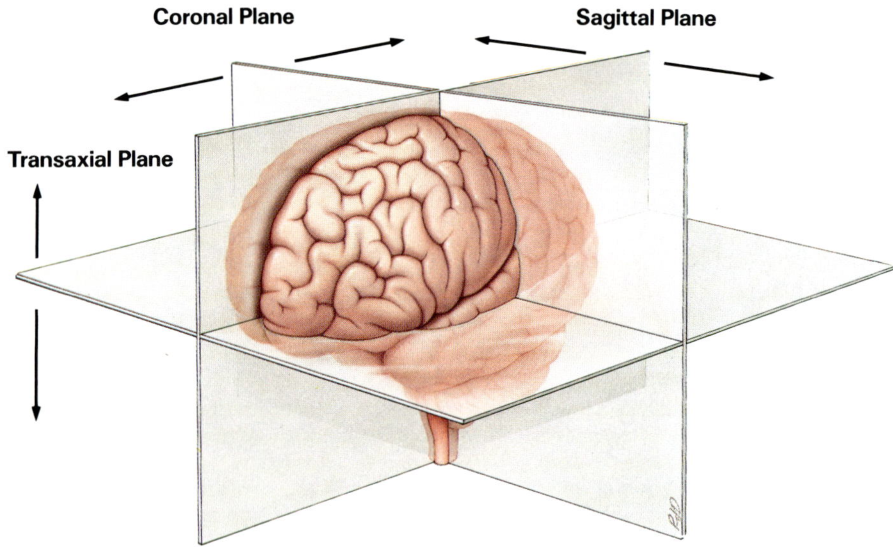

Transaxial Plane

The transaxial cerebral SPECT sections are displayed in the same format as the MRI images. The sections begin inferiorly in the cerebellum and continue superiorly through the entire cerebrum.

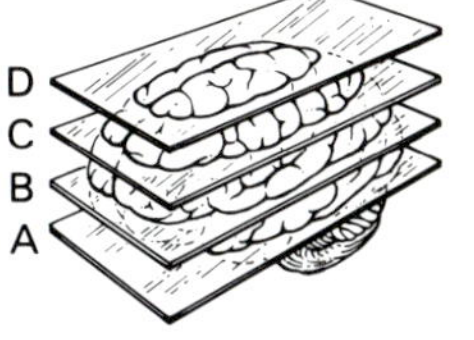

MRI | **IMP/SPECT**

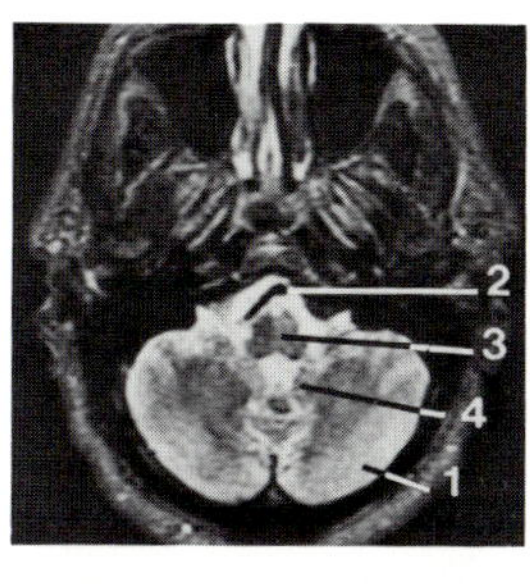

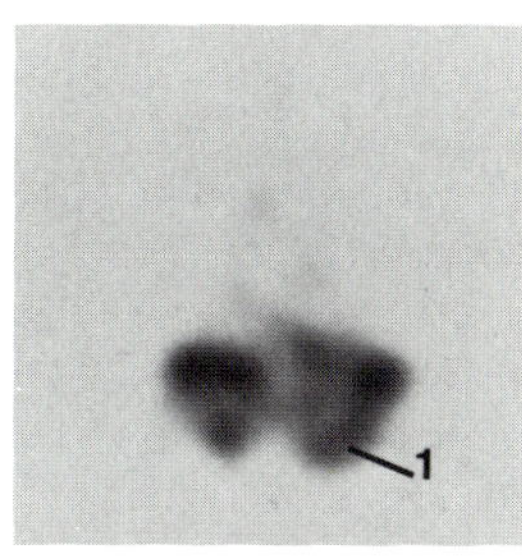

A. 1. Cerebellar hemisphere
2. Vertebral artery
3. Medulla oblongata
4. Cerebellar tonsil

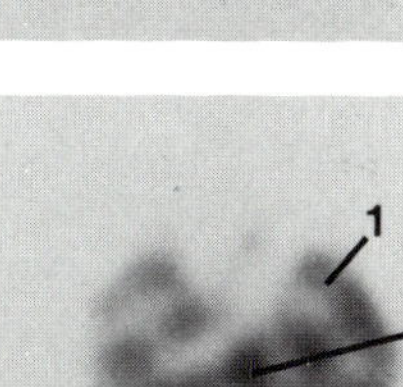

B. 1. Temporal lobe
2. Occipital lobe
3. Upper brain stem
4. Temporal horn
5. Tentorial notch
6. Cerebellar vermis

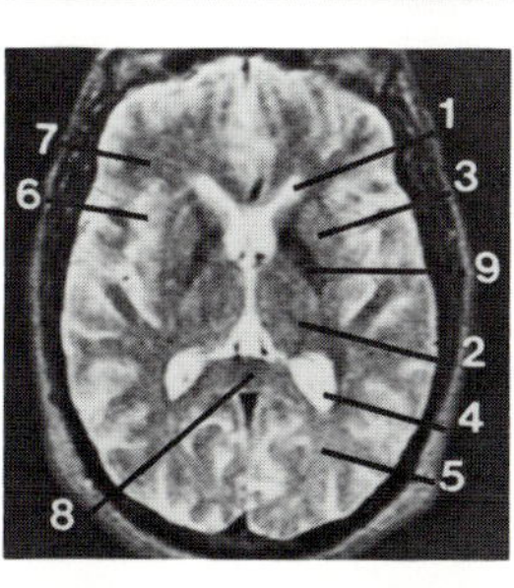

C. 1. Frontal horn
2. Thalamus
3. Basal ganglia-Lentiform nucleus
4. Occipital horn
5. Occipital lobe
6. White matter
7. Frontal lobe
8. Splenium of corpus callosum
9. Internal capsule

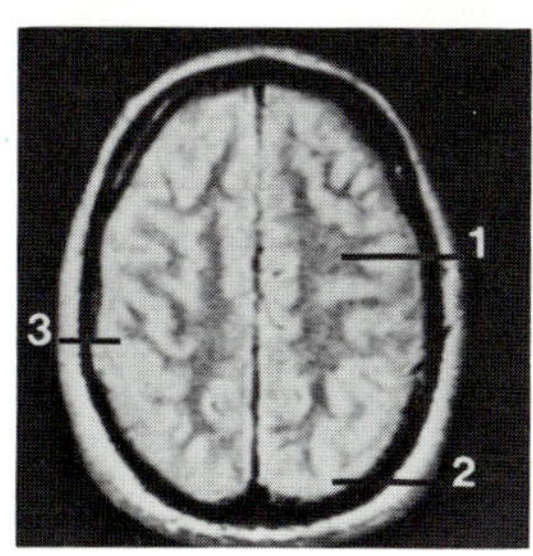

D. 1. White matter
2. Gray matter
3. Parietal lobe

SECTION I: Normal/Correlative Anatomy

Coronal Plane

The coronal cerebral SPECT sections are displayed in the same format as the MRI images. The sections begin anteriorly in the frontal lobes and continue posteriorly through the entire brain.

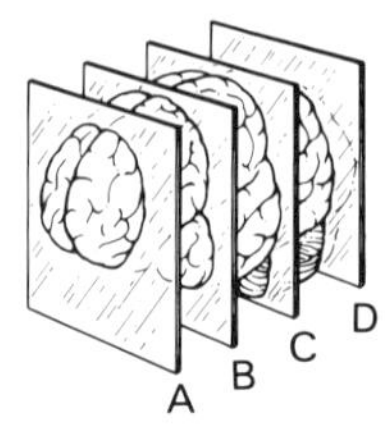

MRI **IMP/SPECT**

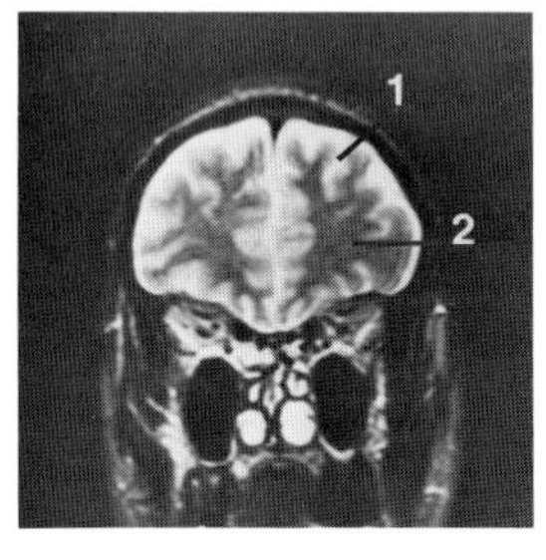

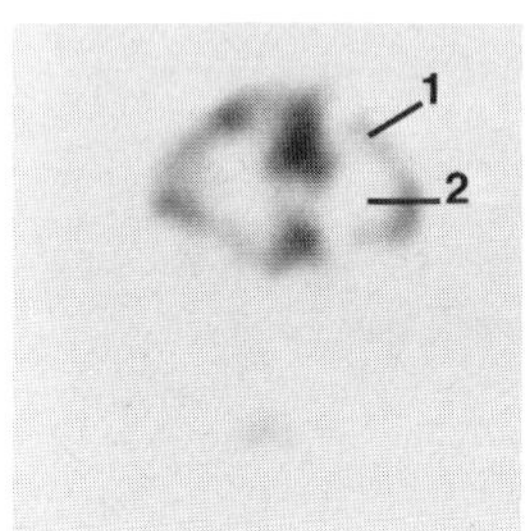

A. 1. Gray matter—Frontal lobe
2. White matter—Frontal lobe

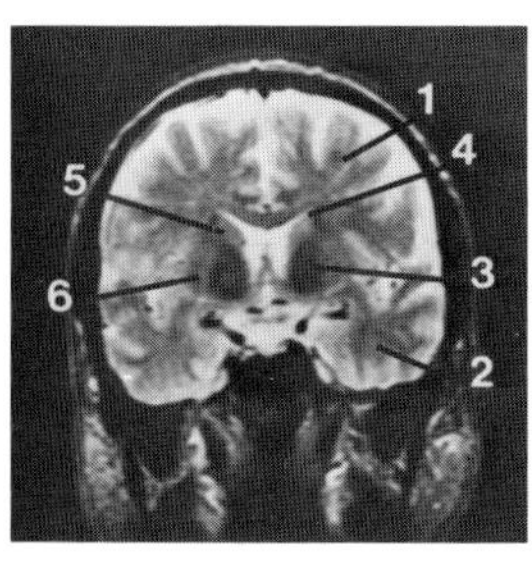

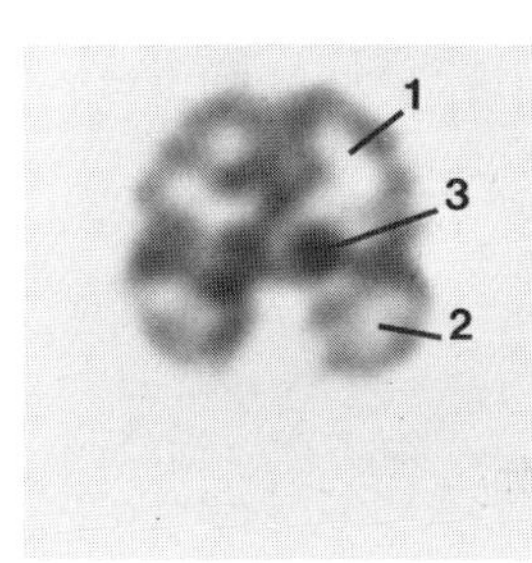

B. 1. White matter—Frontal lobe
2. White matter—Temporal lobe
3. Basal ganglia
4. Frontal horn
5. Head of caudate nucleus
6. Lentiform nucleus

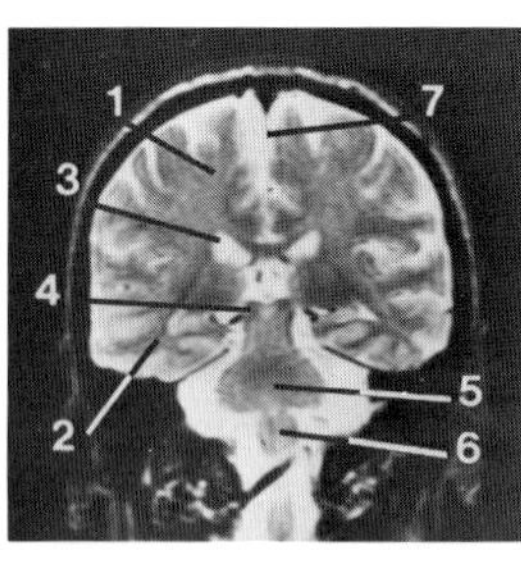

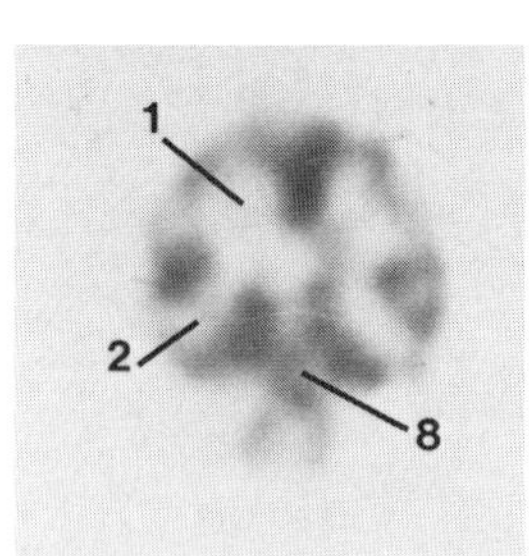

C. 1. White matter—Parietal lobe
2. White matter-Temporal lobe
3. Body of lateral ventricle
4. Mesencephalon
5. Pons
6. Medulla oblongata
7. Interhemispheric fissure
8. Brain stem

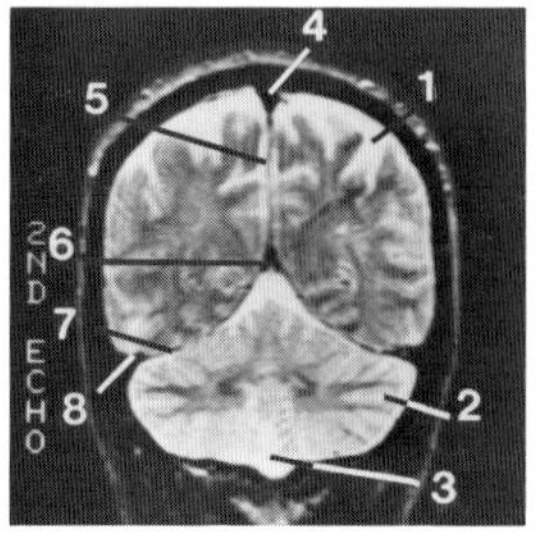

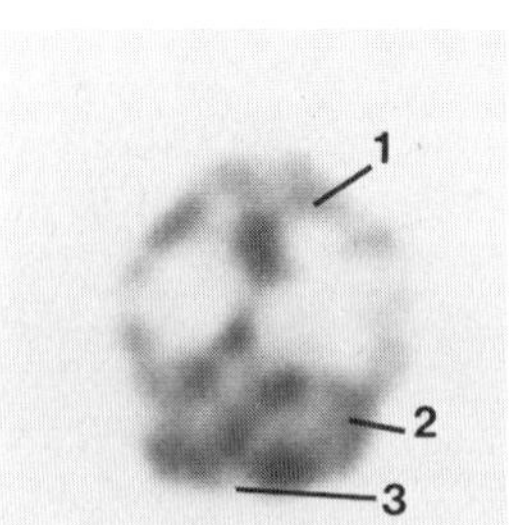

D. 1. Gray matter—Parietal lobe
2. Cerebellar hemisphere
3. Vallecula cerebelli
4. Superior sagittal sinus
5. Falx cerebri
6. Straight sinus
7. Tentorium cerebelli
8. Transverse sinus

Sagittal Plane

The sagittal cerebral SPECT sections are displayed in the same format as the MRI images. The images extend from the right side of the brain to the midline.

A B C D

MRI | **IMP/SPECT**

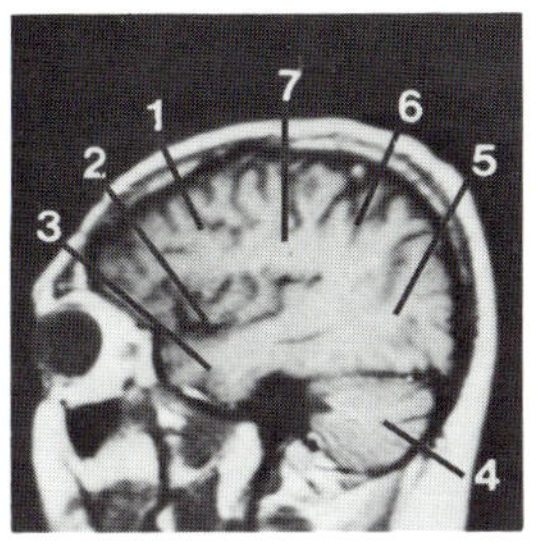

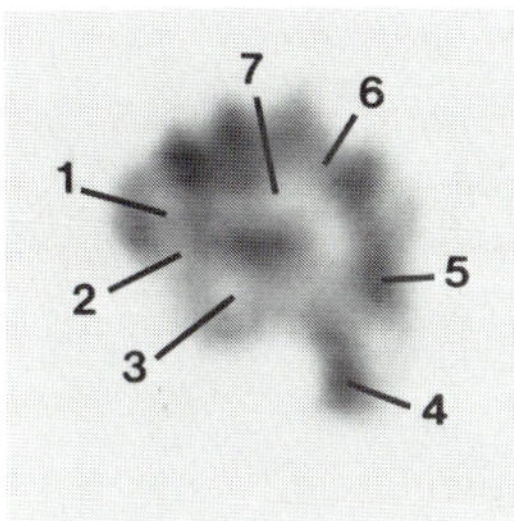

A. 1. Frontal lobe
   2. Sylvian fissure
   3. Temporal lobe
   4. Cerebellum
   5. Occipital lobe
   6. Parietal lobe
   7. White matter

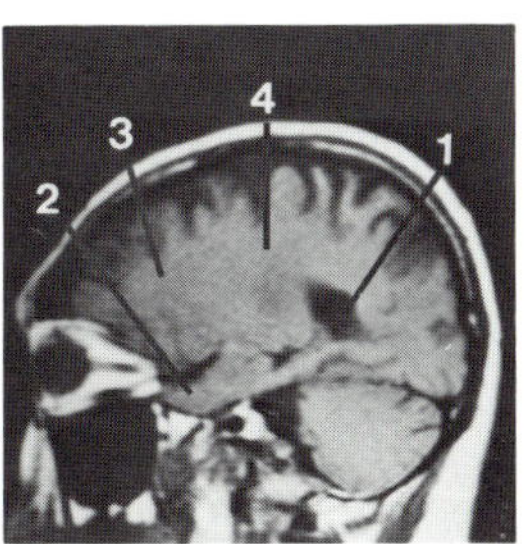

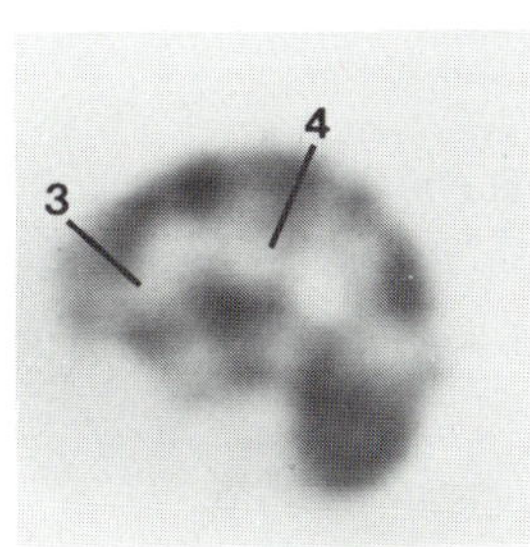

B. 1. Atrium of lateral ventricle
   2. White matter—Temporal lobe
   3. White matter—Frontal lobe
   4. White matter—Parietal lobe

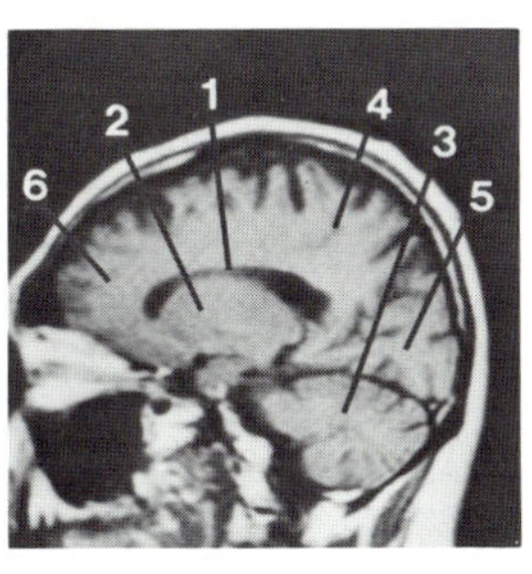

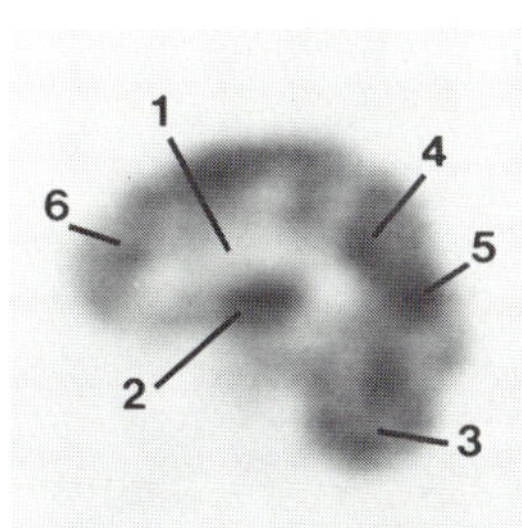

C. 1. Lateral ventricle
   2. Basal ganglia
   3. Cerebellum
   4. Parietal lobe
   5. Occipital lobe
   6. Frontal lobe

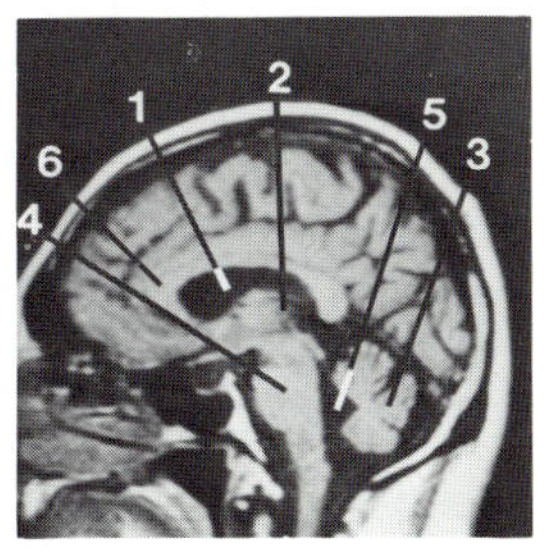

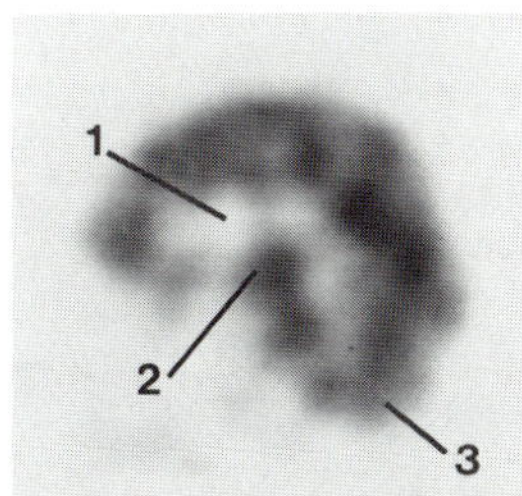

D. 1. Lateral ventricle
   2. Thalamus
   3. Cerebellum
   4. Brain stem
   5. Fourth ventricle
   6. Corpus callosum

INTERPRETING FUNCTIONAL BRAIN IMAGES

Helpful Hints

Normal Tracer Uptake
Typically, there is symmetric distribution of tracer uptake in both hemispheres. Basal ganglia, occipital cortex, and cerebellum will often appear darker than do other regions. Ventricles will not be as sharply defined as with CT or MRI because the tracer is not taken up by the ventricles or by the white matter surrounding the ventricles (Fig. 8).

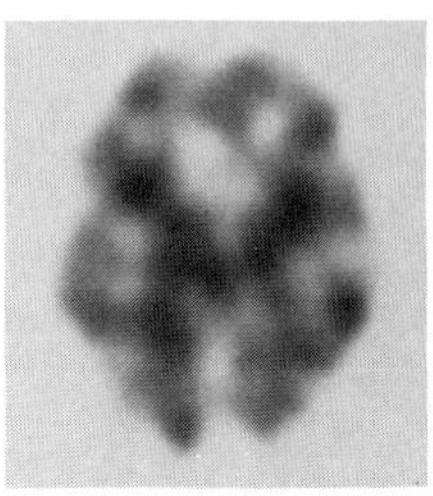

Figure 8. Normal distribution.

Absent Tracer Uptake
Implies that tracer uptake is not present in a zone extending from deep within the brain through the cortical rim. These regions of absent uptake will have the appearance of a bite or wedge taken out of the brain. This type of finding is associated with a region of infarction seen on CT or MRI[6] (Fig. 9).

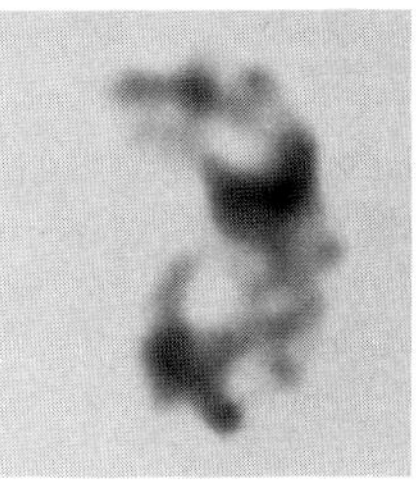

Figure 9. Absent tracer uptake.

Reduced but Not Absent Tracer Uptake
Regions of reduced but *not* absent tracer uptake usually appear "less intense" than do the surrounding cortical regions and corresponding regions of the contralateral hemisphere. They also will appear "thin" or "less intense" in comparison to equivalent regions seen in normal studies. This finding is typically associated with the presence of ischemia and is not necessarily evident on CT or MRI images[7] (Fig. 10).

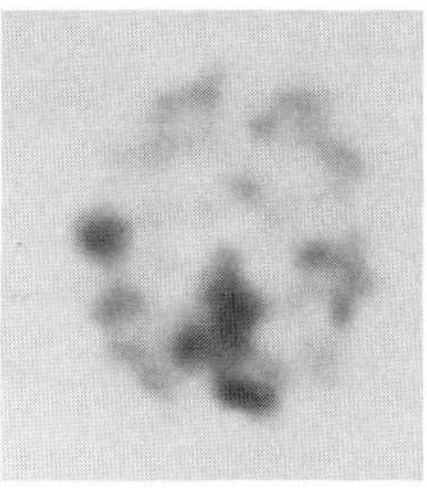

Figure 10. Reduced tracer uptake.

Increased Tracer Uptake
These regions of increased tracer uptake usually appear as having a "greater intensity" than do the surrounding cortical regions and corresponding regions of the contralateral hemisphere. They will also appear "thicker" and "darker" than the equivalent regions seen in normal studies. Findings of increased tracer uptake have been associated with regions of hypermetabolism such as seizure activity[8] (Fig. 11).

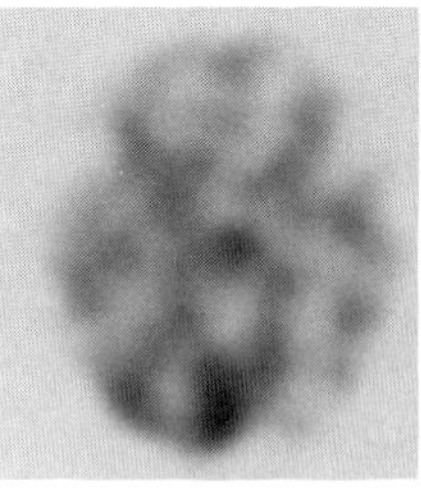

Figure 11. Increased tracer uptake.

SECTION I: C

Pitfalls

PITFALLS

A wide variety of potential pitfalls may be encountered when performing cerebral SPECT imaging studies. The most common pitfalls fall into one of three major groups: patient; technical; and interpretative.

Patient Factors

The major patient factors that may give rise to significant potential imaging pitfalls for most types of SPECT examinations, including cerebral SPECT studies, are: patient size; patient motion; patient position; and low count rates.

Patient size. This is a particular problem in cerebral SPECT imaging, since there may be difficulty in the proper positioning of the patient in the gantry system and also in maintaining the detector head as close to the patient's head as possible. Getting the detector close to the patient's head is of particular concern with obese patients for whom even a cut-off configuration detector may not be able to clear the shoulders and allow an adequately short radius of rotation. The result is a longer-than-desirable radius of rotation(>14 cm), which gives rise to a loss of spatial resolution (Fig. 12).

Patient motion. Movement of the patient's head during the SPECT acquisition will also degrade image quality. Proper quality control during the acquisition procedure will usually detect this abnormality and should be performed on a routine basis (Fig. 13A,B). The technologists should make every effort to keep the patient's head-movement to a minimum.

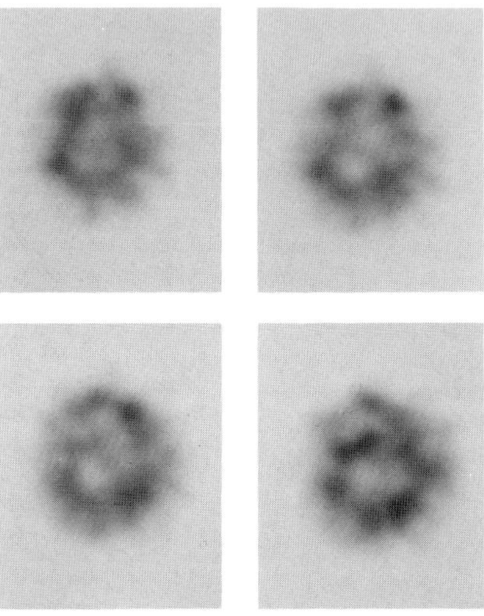

Figure 12. Increased radius of rotation resulting in loss of spatial resolution in a very obese patient.

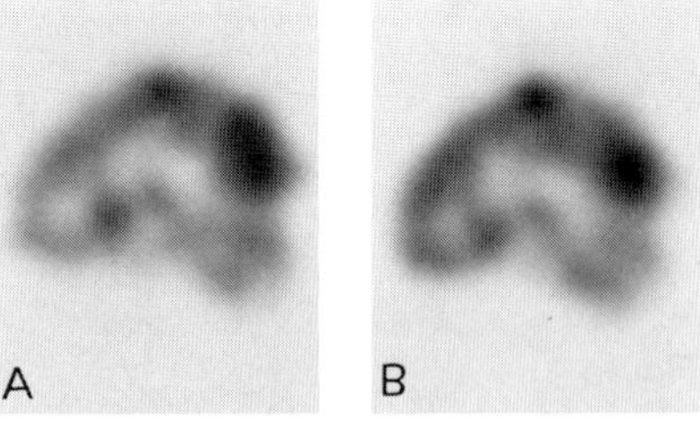

Figure 13.A: Sagittal plane study with motion. B: Sagittal plane study with motion correction and improved spatial resolution.

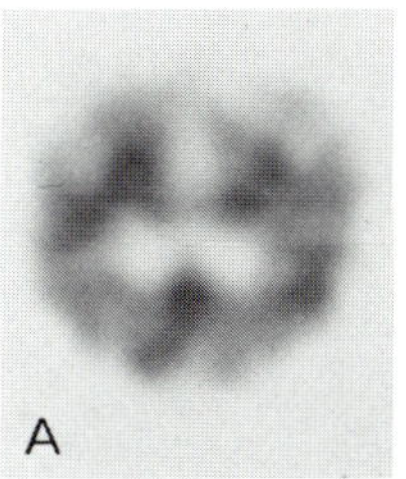

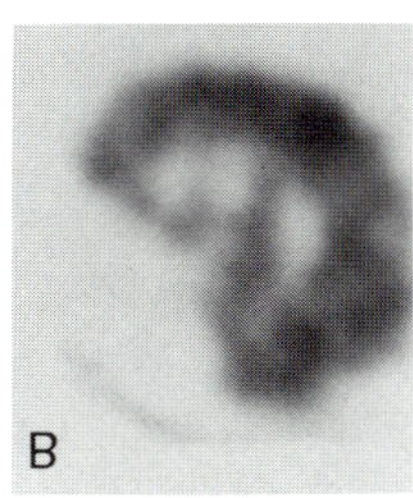

Figure 14. A: Transaxial plane and B: Sagittal plane images with patient in extended neck position causing inadequate delineation of the temporal lobes on the transaxial plane.

Patient position. Improper positioning of the detector in relation to the patient's canthomeatal line can give rise to images that may easily be improperly interpreted (Fig. 14A,B). This type of malpositioning can be corrected by using an oblique angle image reorientation algorithm similar to that currently available from a number of commercial vendors for myocardial SPECT imaging procedures (Fig. 15A,B,C). Although other types of tilting of the patient's head can also be corrected with this approach (Fig. 16A,B,C), not all malalignments can be resolved without doing the study a second time.

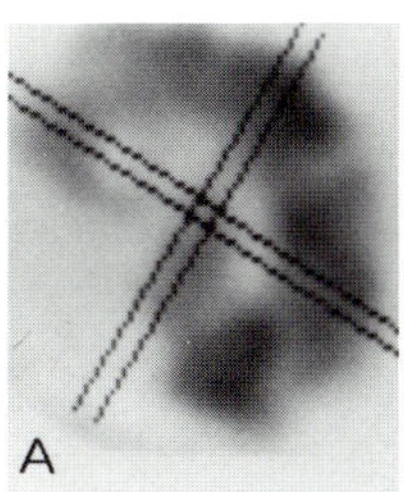

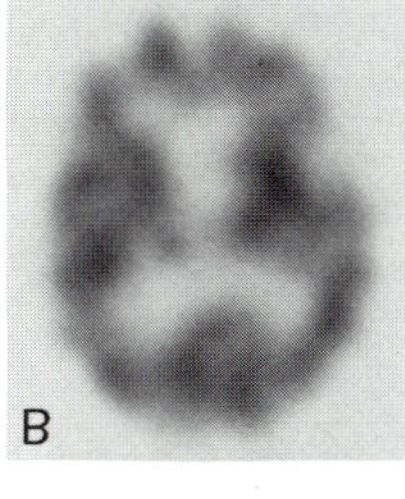

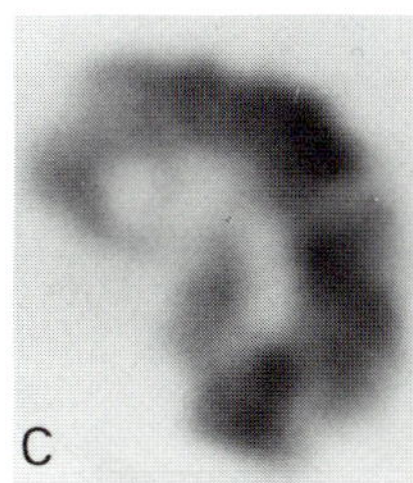

Figure 15: Oblique angle reorientation of images resulting in better delineation of the temporal and frontal lobes. A: Cursor position for oblique angle technique. B: Corrected transaxial plane. C: Corrected sagittal plane.

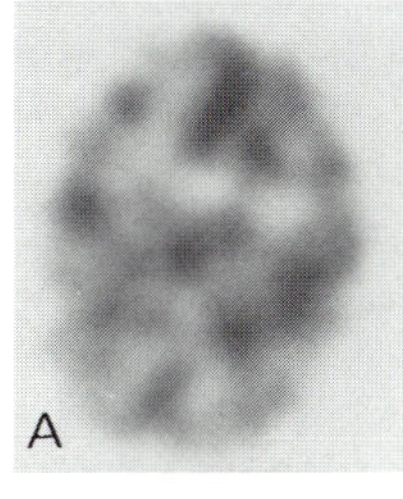

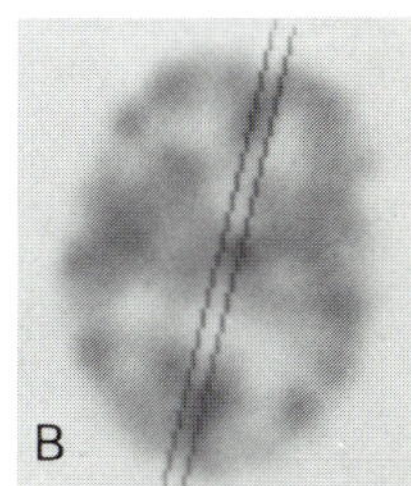

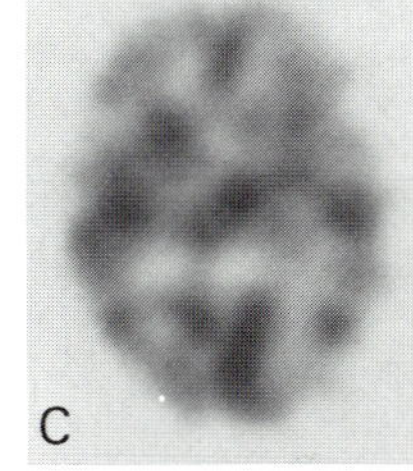

Figure 16. A: Transaxial plane image rotated to patient's left side. B: Cursor placement for oblique angle correction technique. C: Corrected transaxial plane image.

Low count rates. The most frequent cause of a low count rate examination is infiltration of a portion of the injected intravenous dose. The result is usually a noisy reconstruction that may be difficult if not impossible to interpret. The use of a butterfly intravenous injection setup with a 25-gauge needle taped in place before the start of the procedure will reduce the likelihood of an infiltration.

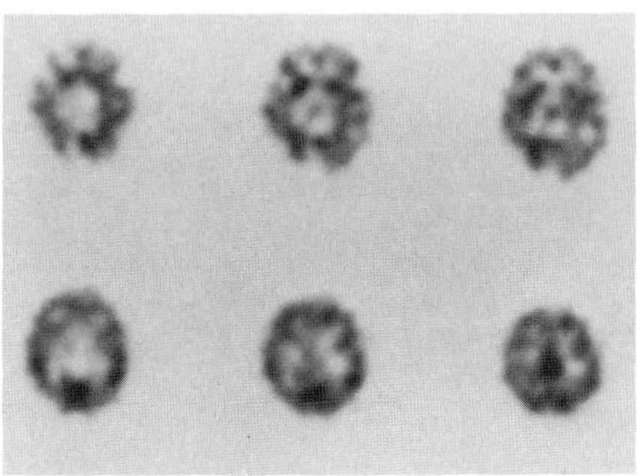

Figure 17. Ring artifact secondary to detector non-uniformity.

Technical Factors

Pitfalls related to technical factors can be categorized into one of three major groups corresponding to the main technical aspects of the imaging procedure: acquisition; processing; and display factors.

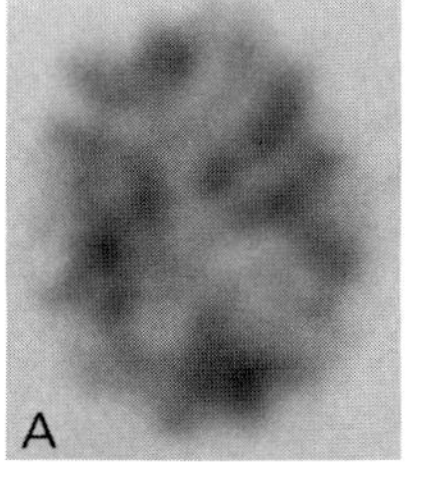

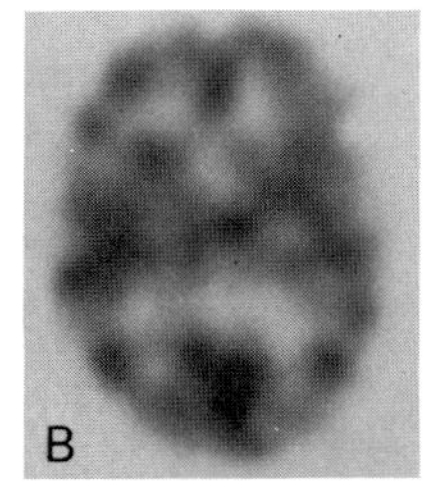

Figure 18. A: Image distortion secondary to center of rotation error. B: Proper center of rotation.

Acquisition factors. Two of the most important instrumentation problems that can cause significant image artifacts during the acquisition procedure are detector nonuniformity and center-of-rotation errors. Detector nonuniformity can give rise to troublesome ring image artifacts that may inadvertently be interpreted as disease (Fig. 17). This type of artifact is avoided by proper and ongoing calibration of the detector photomultiplier tubes and uniformity correction using high count density correction maps, as described in the Instrument Quality Control section of the Appendix. Detector head center-of-rotation errors are another source of artifacts. This type of image distortion becomes increasingly apparent as the size of the center-of-rotation error increases (Fig. 18A,B). This particular problem can be readily identified by routinely performing a center-of-rotation quality control check, as discussed in the Quality Control section of the Appendix. Other acquisition problems that can give rise to potential pitfalls include incomplete acquisition protocols (Fig. 19A,B) as well as inadequate angular sampling (Fig. 20A,B) and spatial sampling.

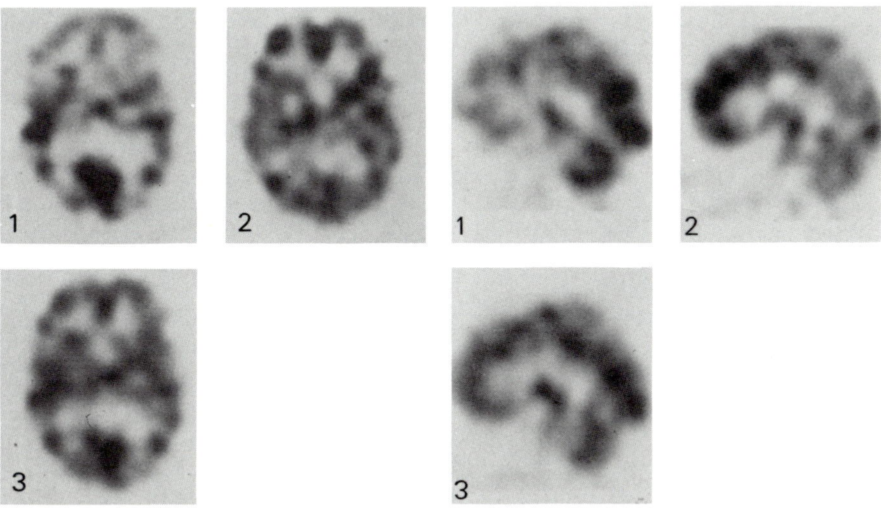

Figure 19. A: Transaxial plane-180° vs 360° acquisition. 1: 180° posterior
2: 180° anterior
3: 360° acquisition.

Figure 19. B: Sagittal plane — 180° vs 360° acquisition. 1: 180° posterior
2: 180° anterior
3: 360° acquisition.

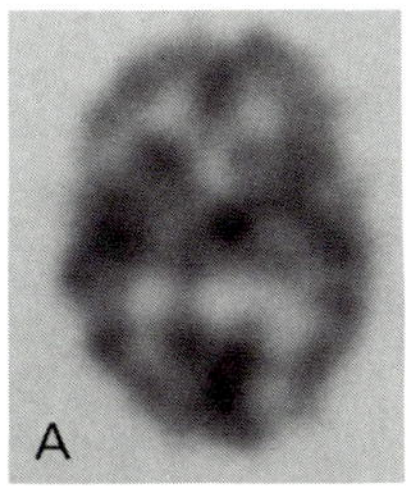

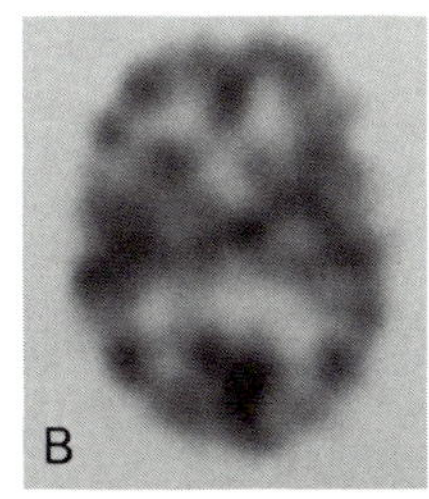

Figure 20. Angular sampling. A: 32 frame. B: 128 frame.

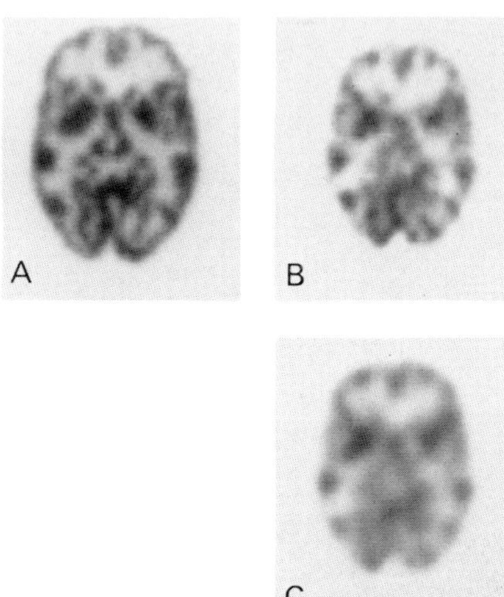

Figure 21. A: High resolution planar image of Hoffman brain phantom. B: Properly filtered transaxial plane SPECT image. C: Over filtered transaxial plane SPECT image with loss of spatial resolution.

Processing factors. Processing problems that can cause significant image degradation usually arise from abusive filtering techniques. Excess filtration of the images will, for example, result in an image that is ''oversmoothed'' (Fig 21A,B,C). Oversmoothed images generally have poor spatial resolution (Fig. 22A,B). On the other hand, inadequate filtration will result in an excessive image noise that can likewise make image interpretation very difficult (Fig. 23A,B). Pixel sizing represents another potential pitfall. Pixel size of the transaxial sections is an area of real concern when reconstructing the reoriented coronal and sagittal planes. The transaxial sections used for reconstruction of the coronal and sagittal planes should be one pixel-width thick (3 to 4 mm)—otherwise, significant image-distortion may result (Fig. 24A,B).

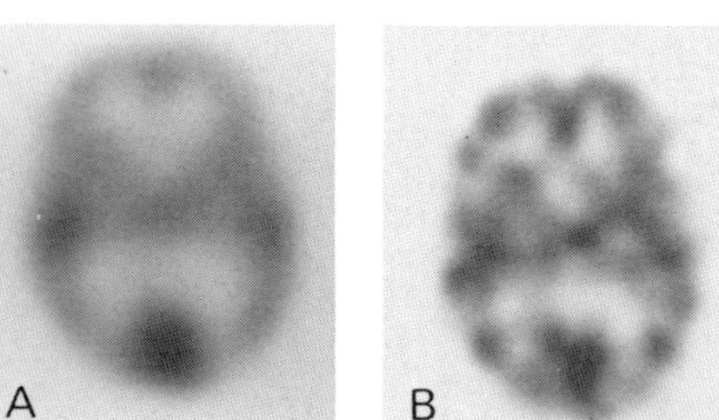

Figure 22. A: Over filtered transaxial plane SPECT image of a patient. B: Properly filtered image of same patient with improved spatial resolution.

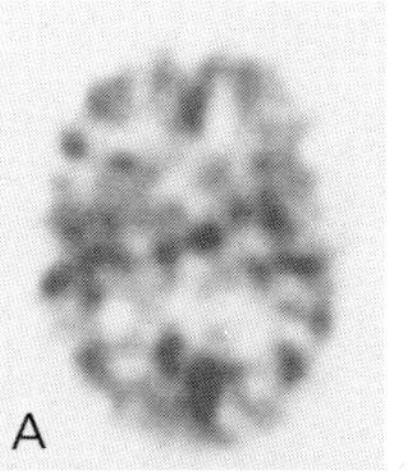

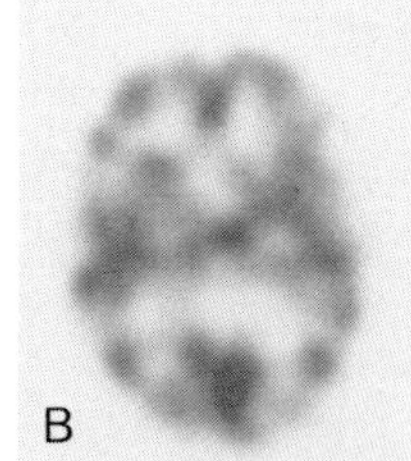

Figure 23. A: Under filtered transaxial plane SPECT image of a patient resulting in an excessively noisy image. B: Properly filtered image of same patient with improved spatial resolution.

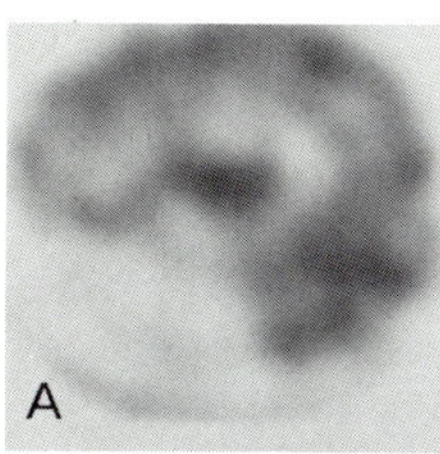

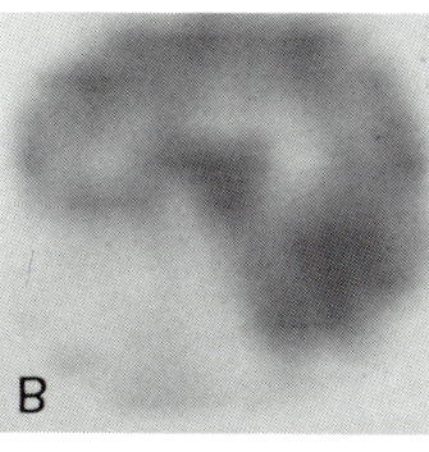

Figure 24. Pixel sizing. A: Distorted sagittal plane secondary to reconstruction from 3 pixel-width transaxial plane study. B: Improved image resolution in sagittal plane study reconstructed from 1 pixel-width transaxial plane study.

Display factors. Another technical area that has the potential to cause significant problems relates to the area of image display. Great care should be given to the proper choice of transparency film used to archive the cerebral SPECT imaging study. Consideration must also be given to the type of black-and-white and color maps used to display the data on the computer monitor and the type of map utilized for the image formatter display. These variables should be properly established when first initiating cerebral SPECT imaging studies and then checked on an ongoing basis. Once these factors are carefully controlled and the film formatters are properly maintained, the most frequent cause for display problems relates to the use of an improper display intensity. Image displays that are either too light or too dark may result in the loss of important data that may be pertinent to the study (Fig. 25A,B,C).

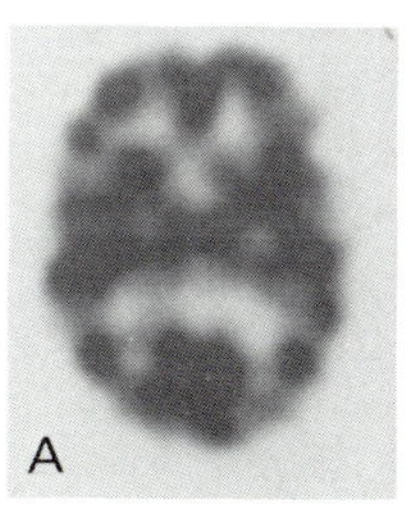

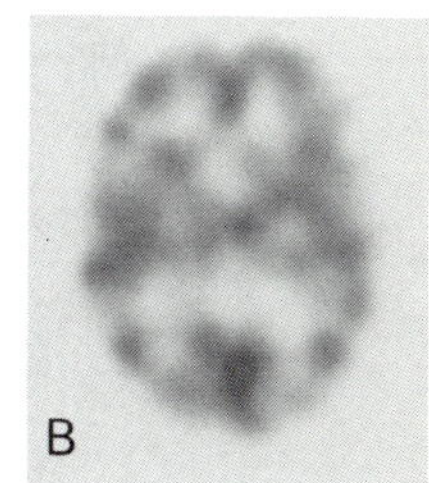

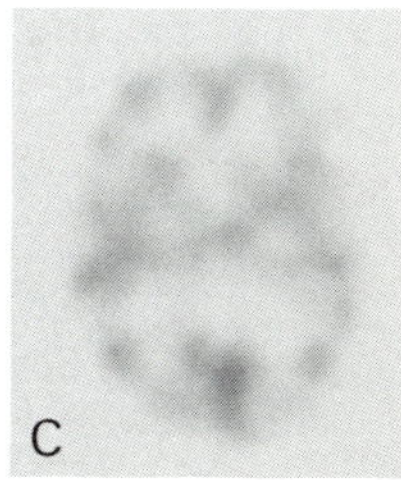

Figure 25. Image display intensity. A: Excessively dark image. B: Proper intensity. C: Excessively light image.

Interpretive Factors

Most of the pitfalls described above can produce images with artifacts that either simulate pathology or make image interpretation very difficult or impossible. Proper quality-control and standardization of the imaging technique should significantly reduce interpretive errors. A strong knowledge of the previously discussed pitfalls, should they occur, will also result in a further reduction in the number of misinterpretations of cerebral SPECT imaging studies. In addition, it is essential that the physician interpreting the cerebral SPECT images have a strong fundamental knowledge of cross-sectional anatomy of the brain in all three orthogonal planes, plus a firm understanding of the major normal variants in anatomic structures.

REFERENCES

1. Hellman RS, Tikofsky RS, Collier BD, Joestgen TM, eds: A new era in functional brain imaging in the evaluation of stroke: A monograph. New York; Medi-Physics: 1988.
2. Hellman RS, Collier BD: Single photon emission computed tomography: A clinical experience. In *Nuclear Medicine Annual 1987.* Freeman LM, Weissmann HS, eds. New York; Raven Press; 51-101; 1987.
3. Polak JF, English RJ, Holman BL: Performance of collimators used for tomographic imaging of I-123 contaminated with I-124. *J Nucl Med* 24: 1065-1069; 1983.
4. Larson SA, Bergstrand G, Berstedt H, et al: A special cut-off gamma camera for high resolution SPECT of the head. *J Nucl Med* 25: 1023-1030; 1984.
5. Woronowicz EM, Eisner RL, Gullberg DJ, et al: Factors affecting single photon emission computed tomography image quality and recommended QC procedures. *General Electric Medical Systems Operations.* Milwaukee; General Electric Company: 1982.
6. Hill TC, Holman BL, Lovett R, et al: Initial experience with SPECT (single photon emission computerized tomography) of the brain using N-isopropyl I-123 p-iodoamphetamine: concise communication. *J Nucl Med* 23: 191-195; 1982.
7. Holman BL, Zimmerman RE, Schapiro JR, et al: Biodistribution and dosimetry of N-isopropyl-p-[^{123}I] iodoamphetamine in the primate. *J Nucl Med* 24: 922-931; 1983.
8. Podreka I, Höll K, Dal Biamco P, et al: Clinical relevance of N-isopropyl-(^{123} I) p-iodoamphetamine (IMP) SPECT brain imaging. In *Amphetamines and pH Shift Agents for Brain Imaging.* Biersack HJ, Winkler C, eds. Berlin; Walter de Gruyter and Company; 109-125; 1986.

SECTION II: A

Cerebrovascular Disease

INTRODUCTION

SPECT is of considerable clinical value in stroke because it provides a three-dimensional representation of perfusion that is neither invasive nor expensive. It requires low dosages of radiation, it is widely available, and it uses conventional nuclear medicine equipment. In cerebrovascular disease, SPECT is useful both in patients with transient ischemia and with completed stroke. In acute cerebral infarction, SPECT imaging demonstrates hypoperfusion within the first two days after onset, a period when CT scans are usually still normal (Case 1). In the acute period, used in conjunction with CT scans, SPECT is helpful in distinguishing between frank infarction and reversible ischemia. In patients with chronic symptoms and signs of cerebrovascular disease but a negative CT, a positive SPECT study implicates cerebral ischemia and suggests the presence of a hemodynamically significant arterial stenosis or occlusion (Cases 2 and 3). SPECT also has a clinical application in monitoring ischemia associated with a vasospasm following subarachnoid hemorrhage (Cases 4 and 5). This is of great importance in planning surgical management. SPECT imaging therefore complements the information available from CT, MRI, and cerebral angiography. By showing both brain perfusion and brain metabolism (on delayed scans), the technique holds great promise in delineating the physiological evolution of acute infarction and in identifying areas of viable but ischemic brain tissue surrounding chronic infarcts (Case 6). Its role in the planning of therapy in acute stroke and in determining prognosis and treatment in the chronic stage will grow in the near future.

FUNCTIONAL LOCALIZATION

Functional localization of strokes and of the reduction of blood flow to the cerebral hemisphere depends on which of the three major arteries are involved and on the laterality of the stroke and the reduced flow.

Anterior Cerebral Artery Infarction (Cases 7-8)

Occlusion of the anterior cerebral artery produces deficits related to destruction of the medial frontoparietal area and the corpus callosum. Infarction or ischemia of the paracentral lobule results in weakness and sensory loss in the contralateral leg. There may be forced grasping and groping of the contralateral hand. When the infarction extends to the upper convexity, there may be proximal arm weakness. When the proximal deep branches supplying the anterior limb and genu of the internal capsule are involved, weakness of the contralateral face and hand are present. Urinary incontinence—and, to a lesser extent, fecal incontinence—may be found with a unilateral or bilateral anterior cerebral artery stoke. Unilateral occlusion of the left anterior cerebral artery results in a language disturbance in which there occurs marked reduction of spontaneous speech, sometimes to the point of muteness. Infarction of the corpus callosum may produce an interhemispheric disconnection syndrome with apraxia and agraphia of the left hand, with an inability to name objects palpated by the left hand without the aid of vision. With bilateral medial frontal lobe infarction, there occurs a state of akinetic mutism in which there is unresponsiveness to the environment in the absence of alterations of the sensorimotor mechanism or coma.

Left Middle Cerebral Artery Infarction (Cases 9-14)

Total infarction in the territory of the left middle cerebral artery produces global aphasia with a right

hemiparesis and right hemisensory loss, often with right hemianopia and poor conjugate gaze to the right in the acute state. Involvement of the superior division of the left middle cerebral artery produces a frontocentral parietal infarct with Broca's aphasia, hemiparesis, and apraxia of the unparalized left upper extremity. Broca's aphasia is characterized by relatively intact auditory comprehension, with a severe reduction in the amount of grammatical complexity of speech. Involvement of the inferior division of the left middle cerebral artery produces temporal lobe infarction, with associated Wernicke's aphasia but without motor or sensory deficits. Right hemianopia—or, more commonly, a superior quadrantanopia—is sometimes found. Wernicke's aphasia is characterized by poor auditory comprehension and copius fluent speech that contains little meaning. Focal involvement of the left parietal lobe produces agraphia, alexia, and aphasia in which the ability to name objects is the major deficit.

Right Middle Cerebral Artery Infarction (Cases 15-16)

Occlusion of the right middle cerebral artery, in addition to producing left hemiparesis and hemisensory deficit, results in sensory neglect of the left visual space, difficulty drawing and copying, left visual field deficit, and poor conjugate gaze to the right in the acute state. The area of the right hemisphere most apt to produce sensory neglect is the parietal lobe. Patients with severe sensory neglect may manifest anosognosia, a condition in which the patient denies the sensory or motor deficit. Deep, small lesions in the territory of the lenticulostriate arteries take the form of lacunar strokes that—because of the involvement of the internal capsule—produce varying degrees of hemiparesis of the contralateral side without major abnormalities of higher cortical function.

Posterior Cerebral Artery (Cases 17-19)

Occlusion of the left posterior cerebral artery results in infarction of the mesial occipital lobe and produces a right visual field defect, sometimes in association with alexia but without a concomitant writing disorder or aphasia. Involvement of the right posterior cerebral artery produces a left visual field deficit sometimes accompanied by posopagnosia, the impaired ability to recognize familiar faces. Bilateral posterior cerebral artery lesions produce cortical blindness, often in association with a severe memory disorder because of the involvement of inferior medial temporal lobe branches that supply left and right hippocampi. When the penetrating branches of the posterior cerebral artery are involved, thalamic infarction may occur, with associated numbness or decreased sensibility of the opposite side of the body. Some patients with left thalamic strokes develop transient aphasia. Strokes may also involve the posterior fossa area including the cerebellar hemispheres (Case 20).

CASE 1: SPECT studies acquired on GE 400AC/T; STAR II.

This 76-year-old woman was referred for evaluation of an acute infarction of the left cerebral hemisphere. Her history was significant for an infarction of her right cerebral hemisphere a year earlier.

A CT scan (Fig. 1.1) showed hypodensities in the right frontal and temporal lobes and basal ganglia, consistent with a previous infarction of the right middle cerebral artery. There was no definite abnormality in the left cerebral hemisphere.

The cerebral SPECT study (Fig. 1.2) in the transaxial (A) and coronal (B) planes revealed an absence of tracer uptake in a large area of the left frontal, temporal, and parietal lobes, a portion of the occipital lobe, and the left basal ganglia and thalamus. On the right side, there was no tracer uptake in the right frontal lobe, the anterior parietal lobe, the basal ganglia, and the thalamus.

Teaching Points:
1. The pattern of acute symptoms, the large, left-sided defect on the cerebral SPECT study (associated with the lack of a definitive abnormality of the left hemisphere on CT scan) is most consistent with an acute infarction of the left cerebral hemisphere.
2. A very large defect of the cerebral hemisphere is most likely the result of an occlusion of the intenal carotid artery on the involved side.

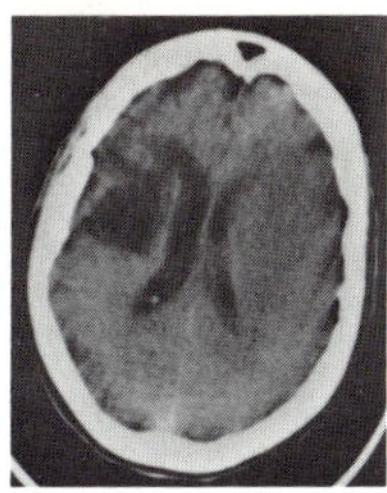

1.1. CT scan.

1.2. SPECT study. A: Transaxial and B: Coronal planes.

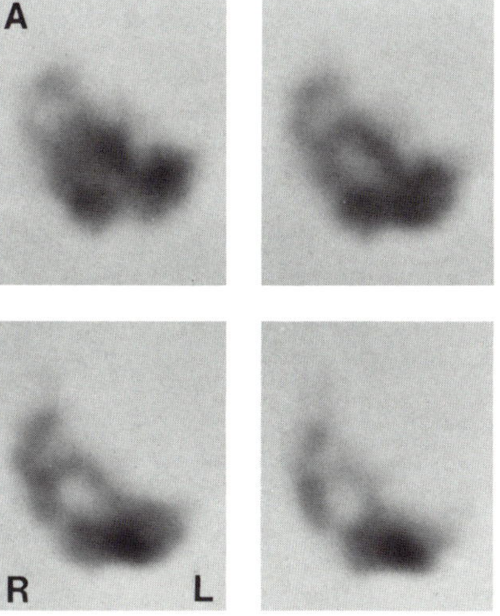

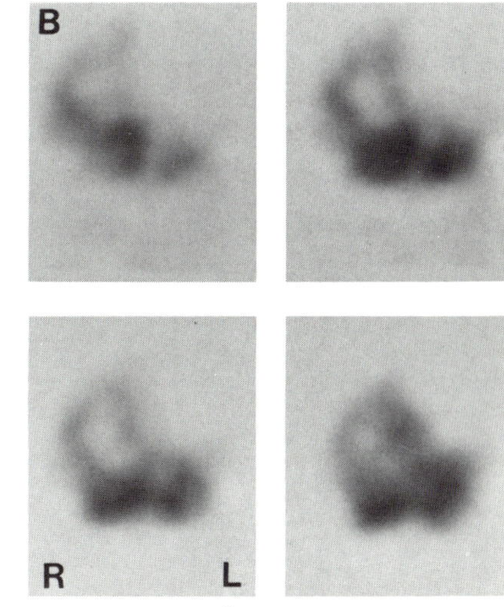

CASE 2: Contributed by Alan H. Maurer, M.D., SPECT studies acquired on GE STARCAM.

This 64-year-old man presented with an episode of predominantly left upper-extremity weakness and an occasional episode of blurred vision.

A CT scan (Fig. 2.1) showed evidence of an old lacunar infarction in the region of the right external capsule.

An angiogram of the aortic arch (Fig. 2.2) revealed a 50% stenosis, with an irregular and probably ulcerated plaque of the right common carotid artery (large arrow) and a shallow, irregular plaque of the innominate (small arrow) and left subclavian arteries.

Immediate and four-hour-delay cerebral SPECT images (Fig. 2.3) in the transaxial plane (A) demonstrated decreased tracer uptake, predominantly in the right posterior watershed area (double arrows), which filled in on the four-hour-delay images (arrow). The coronal (B) and sagittal plane (C) images also showed decreased tracer uptake in the right cerebral hemisphere, primarily in the posterior parietal—occipital regions on the early phase (double arrows), which filled in on the delayed images. Absent tracer uptake was also evident in the right basal ganglia (arrow), which remains fixed on the delayed phase. The overall findings are consistent with an ischemic pattern in the right cerebral hemisphere and an infarction of the right basal ganglia.

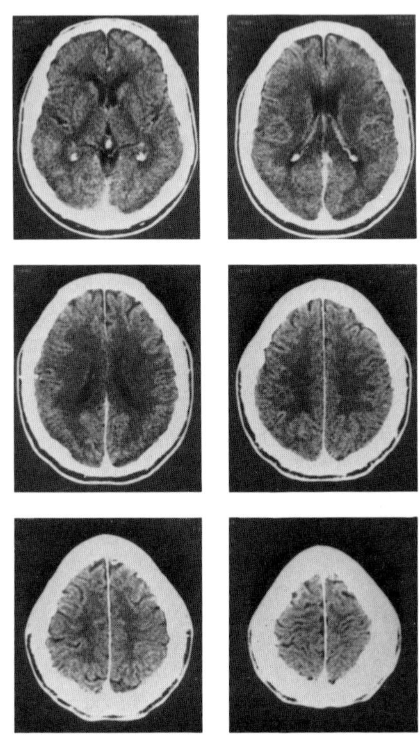

2.1. CT scan.

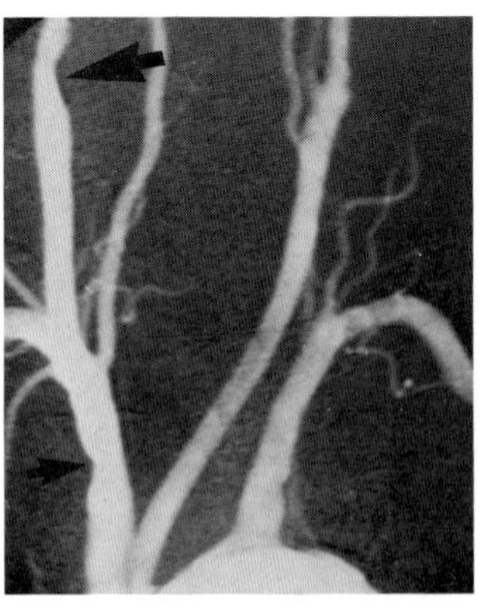

2.2. Angiogram.

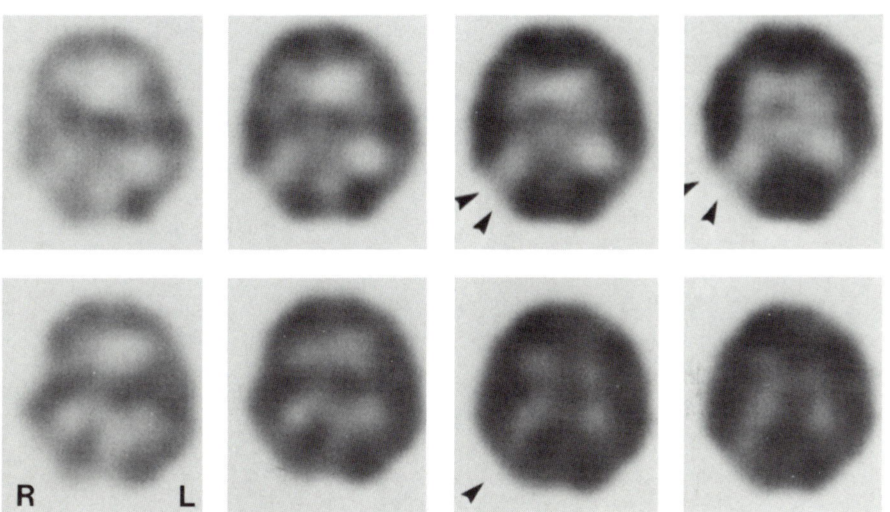

2.3. A: Immediate (upper row) and delayed (lower row) SPECT study in transaxial plane.

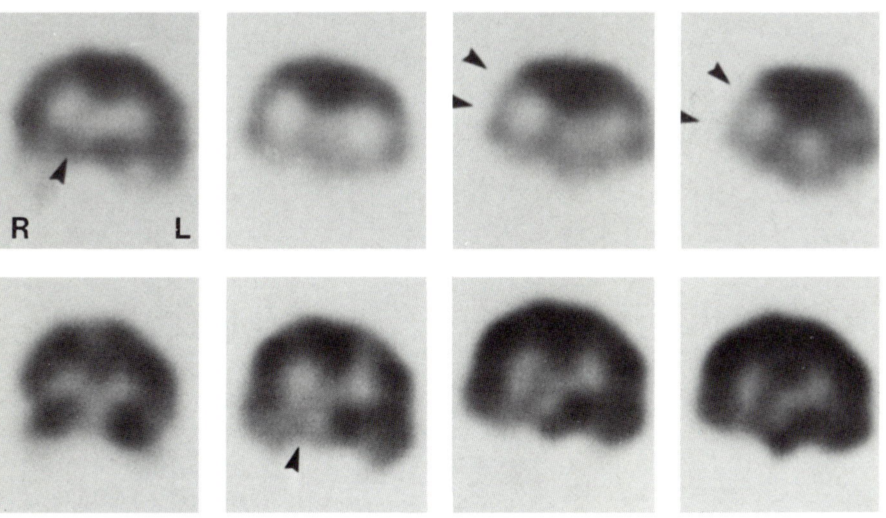

2.3 B: Immediate (upper row) and delayed (lower row) SPECT study in coronal plane.

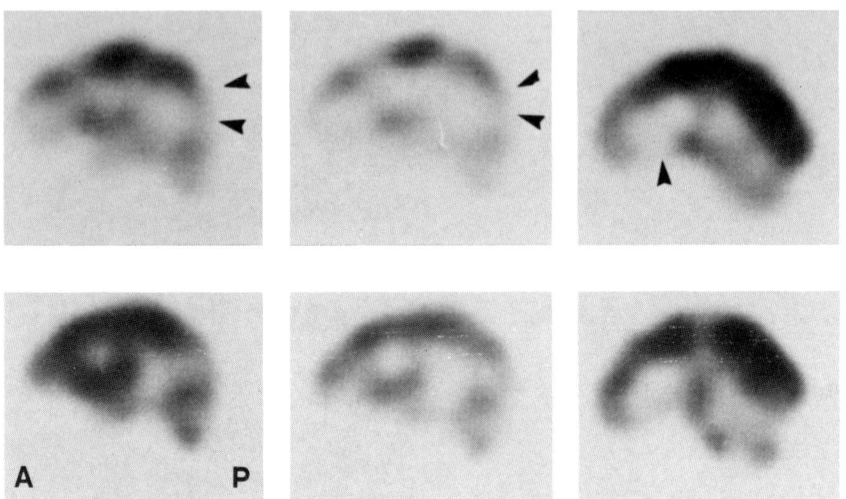

2.3 C: Immediate (upper row) and delayed (lower row) SPECT study in sagittal plane.

CASE 3: *Contributed by Harvey L. Hecht, M.D., SPECT studies acquired on Picker SX 300; PCS Computer.*

This 83-year-old woman was referred for evaluation of gradual, progressive loss of memory over the preceding two years.

A CT scan showed mild cerebral atrophy. The transaxial plane of the cerebral SPECT study (Fig. 3.1) showed decreased tracer uptake in the left posterior parietal-occipital region (arrow heads). This distribution is much more compatible with a vascular etiology of the patient's dementia than is Alzheimer's disease.

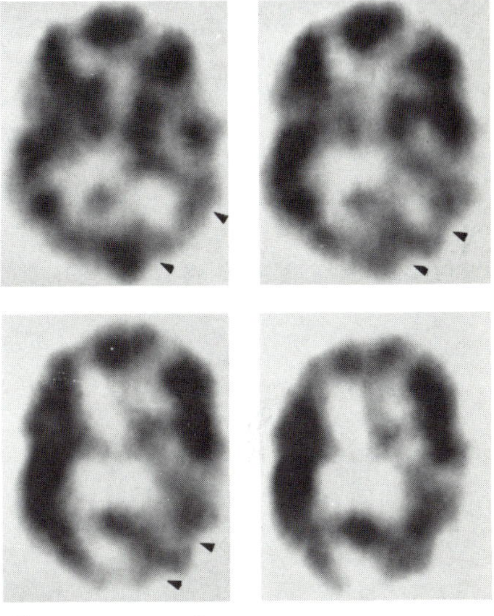

3.1. SPECT study in transaxial plane.

CASE 4: SPECT studies acquired on GE 400AC/T; STAR.

This 38-year-old, right-handed woman was admitted for continued management of a subarachnoid hemorrhage secondary to a basilar tip aneurysm. During the course of her hospitalization, she had intermittent elevations of blood pressure, with peak systolic BPs greater than 200 mm and diastolic BPs of 80 to 110 mm Hg. The patient remained neurologically intact until the fourth day after admission (day 10 since the onset of the hemorrhage). At that time, she appeared progressively confused, and the disorder progressed to a marked aphasia and right hemiparesis.

A CT scan (Fig. 4.1) on the sixth day after admission showed the residual of a subarachnoid hemorrhage, mild ventricular dilatation, and a basilar tip aneurysm (arrow). A follow-up CT scan (Fig. 4.2) on the tenth day after admission revealed a large, posterior parietal-occipital infarction of the left hemisphere and persistent, mild ventricular dilatation. A MRI examination (Fig. 4.3) on the twelfth day after admission showed a left cerebral infarction predominently involving the parietal lobe.

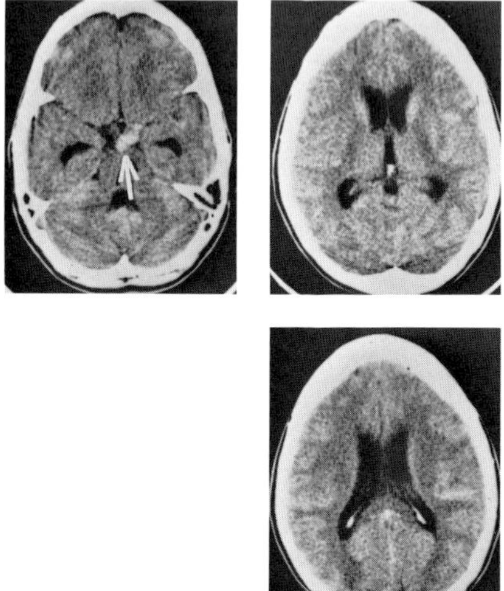

4.1. Initial CT scan.

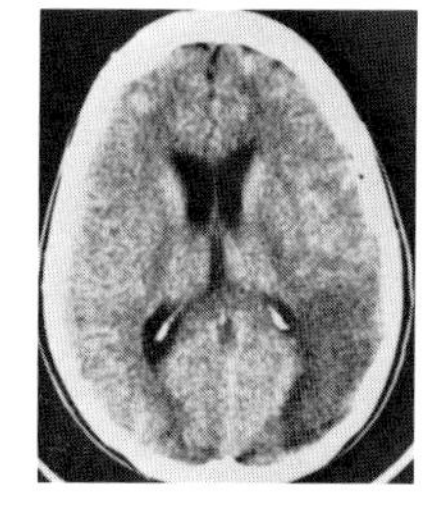

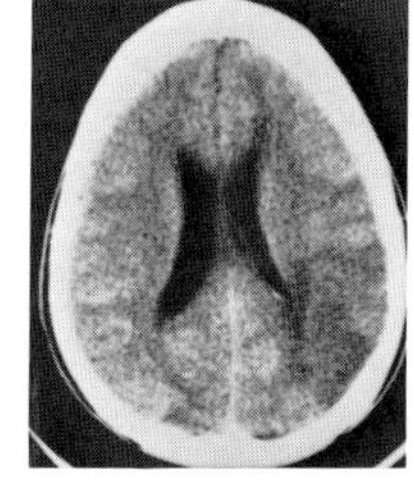

4.2. Follow-up CT scan.

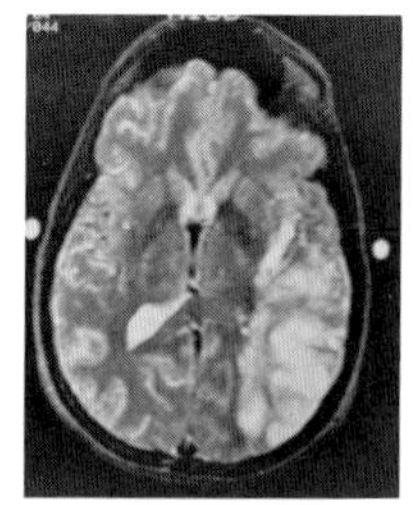

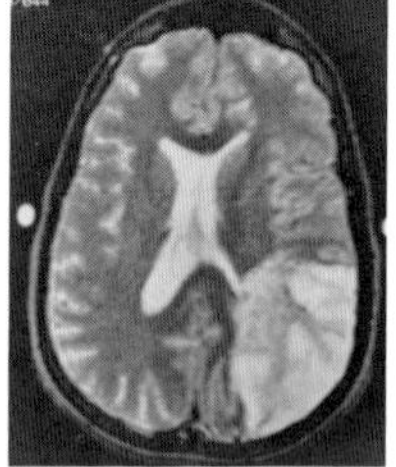

4.3. MRI study.

The initial cerebral SPECT study (Fig. 4.4) in the transaxial plane was done on the same day as the initial CT scan. This study showed decreased tracer deposition in the left posterior parietal-occipital region and in the left posterior temporal lobe (arrow heads), suggesting either infarction or severe ischemia. A second cerebral SPECT study, with both immediate (Fig. 4.5A) and delayed (Fig. 4.5B) phases, was performed four days later. This study showed a "fixed" region (in the left posterior parietal-occipital area) of decreased tracer deposition that was compatible with infarction (arrow heads). In addition, decreased tracer deposition was noted throughout the remainder of the left hemisphere, with improved tracer deposition on the delayed images (arrows). This latter finding was felt to be compatible with extensive ischemia.

Teaching Point:
Cerebral SPECT studies may prove to be useful in following patients with subarachnoid hemorrhage.

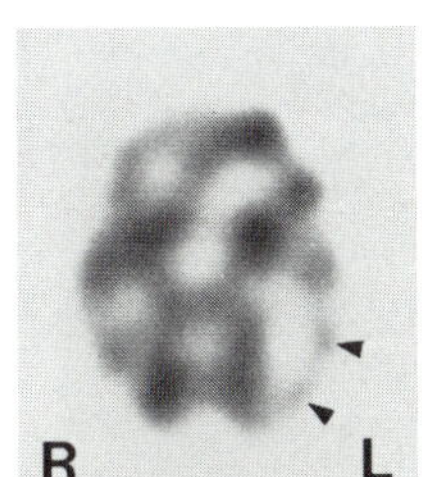

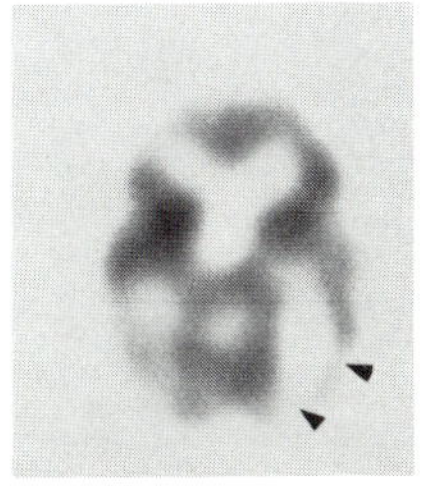
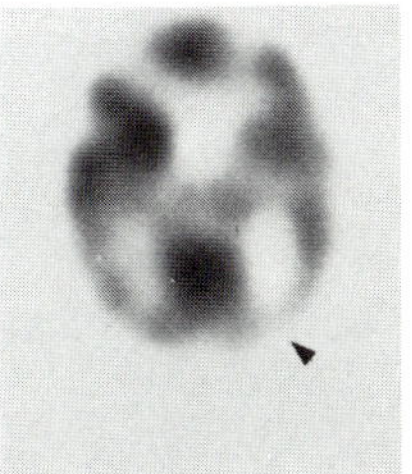
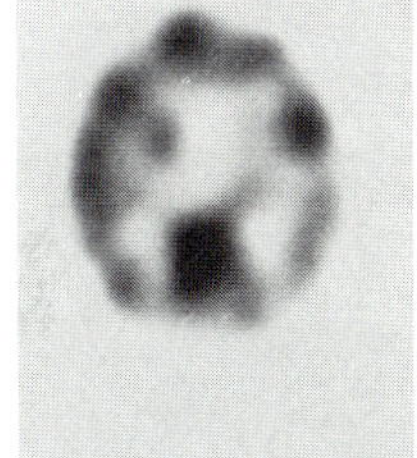

4.4. Initial SPECT study in transaxial plane.

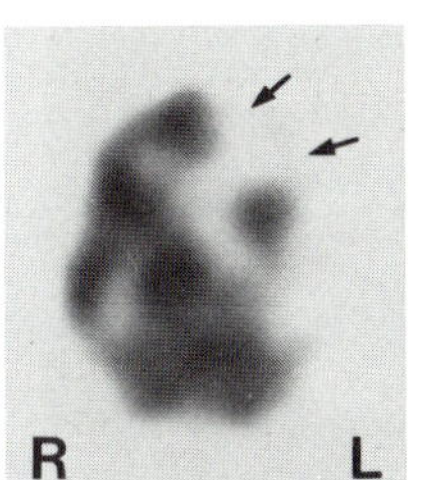

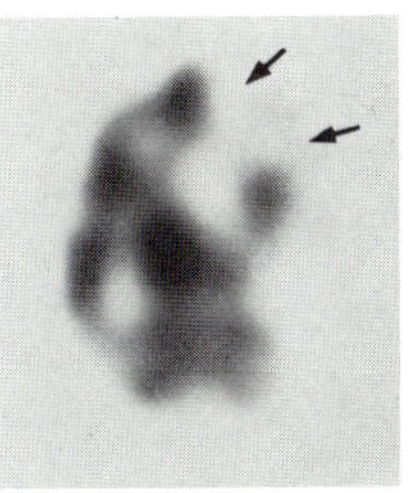
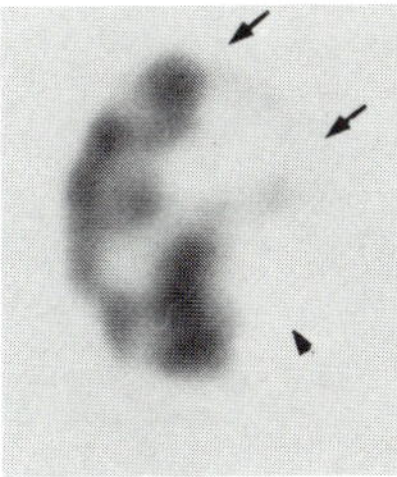
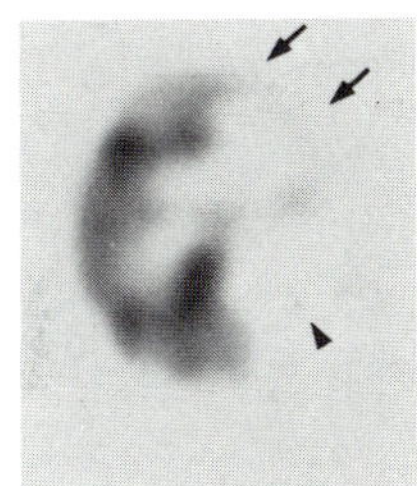

4.5A. Immediate follow-up SPECT study.

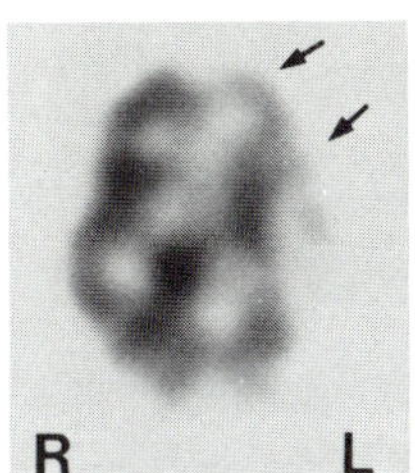

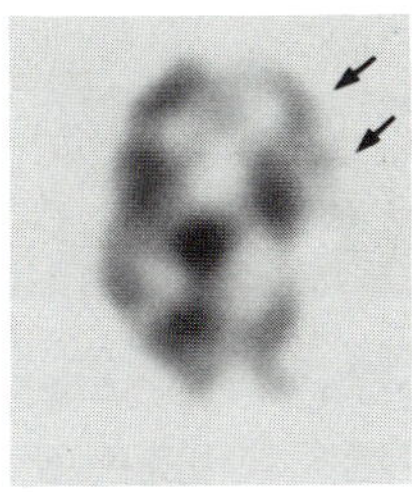
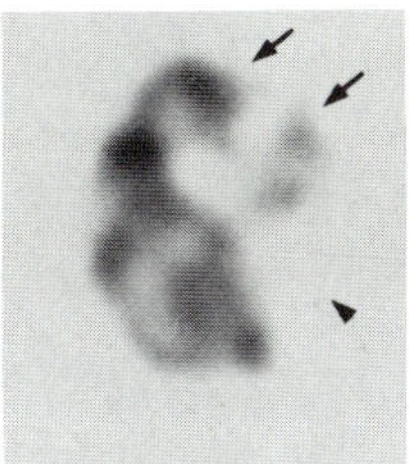
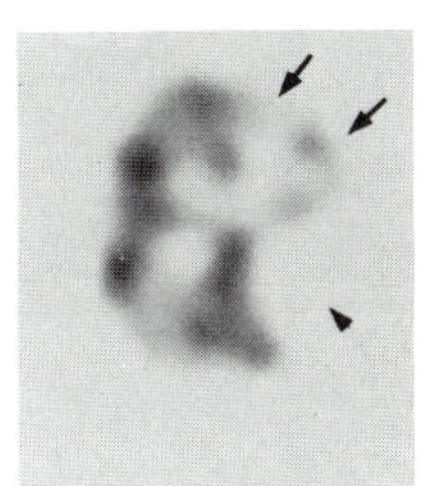

4.5B. Delayed follow-up SPECT study.

CASE 5: *Contributed by Joji Nagawara, M.D., SPECT studies acquired on Shimadzu HEADTOME SET-031.*

This 44-year-old man presented with complaints of headaches, vomiting, and feeling faint. He had a left upper extremity paresis, and subsequent work-up disclosed a ruptured aneurysm arising from the right middle cerebral artery. Following surgical clipping of the aneurysm, the patient's left-sided paresis increased in severity.

Serial cerebral SPECT studies in the transaxial plane (Fig. 5.1) revealed a progressive decrease in the size of a defect in the right middle cerebral artery territory (arrow heads). A residual defect persisted at 90 days. The overall pattern was compatible with combined ischemia (vasospasm) and infarction.

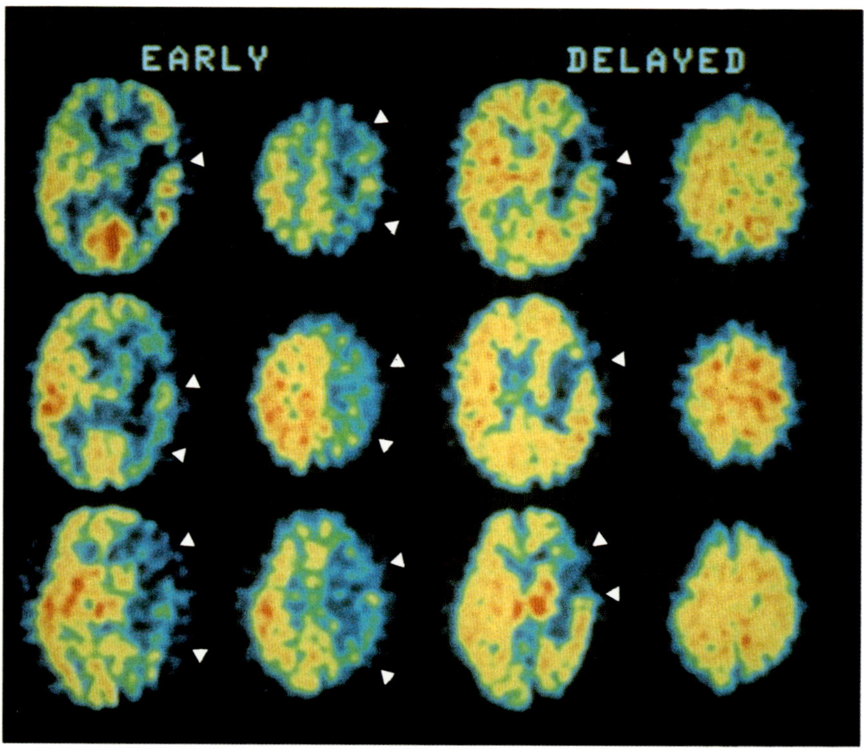

5.1. Serial SPECT study in transaxial plane.

CASE 6: *SPECT studies acquired on GE 400AC/T; STAR.*

This 29-year-old, right-handed woman was referred for evaluation of an episode of generalized weakness, slurring of speech, and blurring of vision associated with a migraine headache. Physical examination revealed a defect of the left visual field. Clinically, the patient was thought to have sustained an infarction of the right occipital lobe, possibly secondary to migrainous vasoconstriction.

A MRI study (Fig. 6.1) confirmed the diagnosis of a right occipital lobe infarction. A cerebral SPECT study in the transaxial plane (Fig. 6.2), performed on the same day as the MRI examination, showed absent tracer deposition in the right occipital lobe (arrow), corresponding to the area of abnormality on MRI. A follow-up cerebral SPECT study (Fig. 6.3), performed two days later, revealed a decrease in size of the abnormal area in the right occipital lobe (arrow). The overall findings were compatible with a combined pattern of infarction with a reversible ischemic zone surrounding the area of infarction.

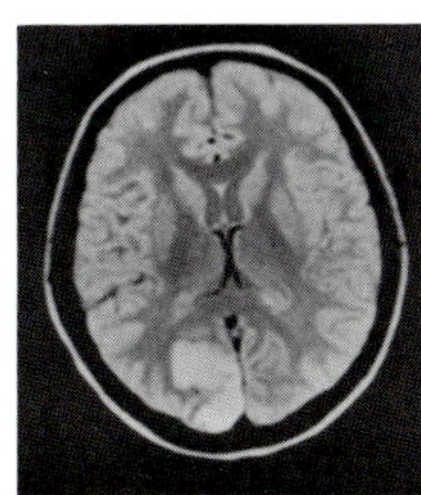

6.1. MRI study.

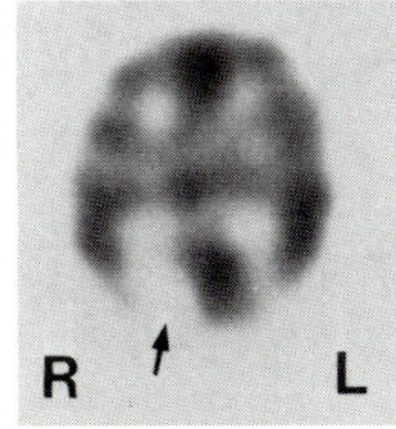

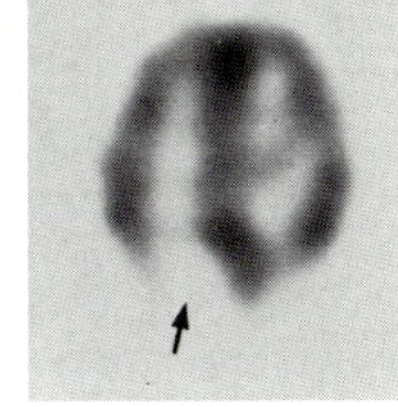

6.2. Initial SPECT study in transaxial plane.

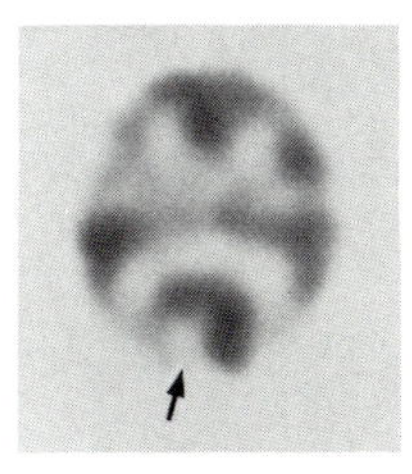

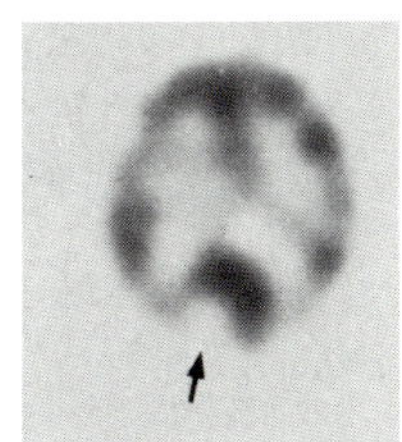

6.3. Follow-up SPECT study.

CASE 7: SPECT studies acquired on GE 400AC/T; STAR.

This 69-year-old man was referred for evaluation of progressive dementia, possibly of the Alzheimer's type. A cerebral SPECT study (Fig. 7.1) in the transaxial (A), coronal (B), and sagittal (C) planes revealed decreased tracer uptake in the medial aspect of the right frontal lobe (arrow). This pattern is typical of an infarction of the anterior cerebral artery.

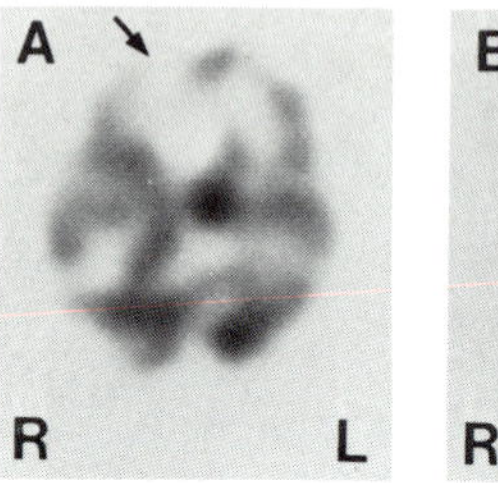

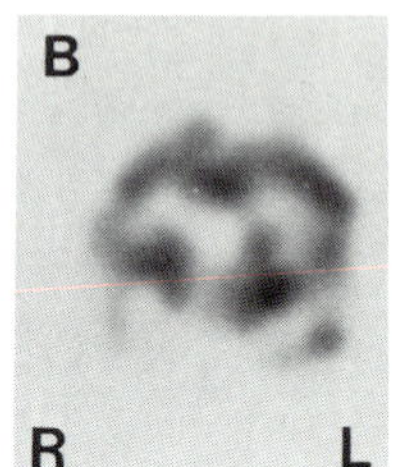

7.1. SPECT study. A: Transaxial; B: Coronal; and C: Sagittal planes.

Teaching Point:
The pattern demonstrated in this case is quite different from the bilateral reduction of tracer uptake in the posterior temporal parietal lobe, which is typically associated with Alzheimer's disease. The pattern is much more compatible with infarction as a cause of the patient's dementia.

__CASE 8:__ Contributed by Harvey L. Hecht, M.D., SPECT studies acquired on Picker SX 300; PCS Computer.

This 71-year-old woman was referred for evaluation of a suspected stroke and an associated seizure disorder.

A CT scan (Fig. 8.1), performed before and after intravenous contrast administration, showed mild dilatation of the lateral ventricles. In addition, there was an area of encephalomalacia and calcification in the left frontal lobe probably due to previous hemorrhage.

The cerebral SPECT study (Fig. 8.2) in the transaxial plane revealed a region of absent tracer deposition in the left frontal lobe (arrow). The SPECT study defect is larger than the region of abnormality on CT.

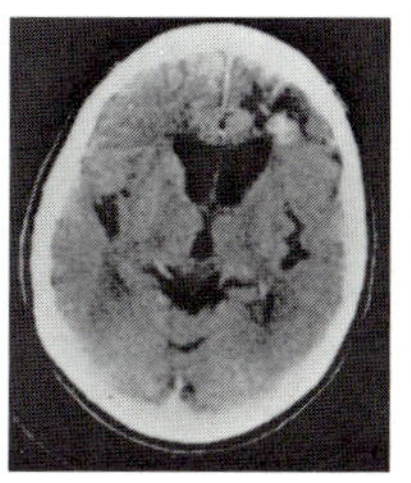

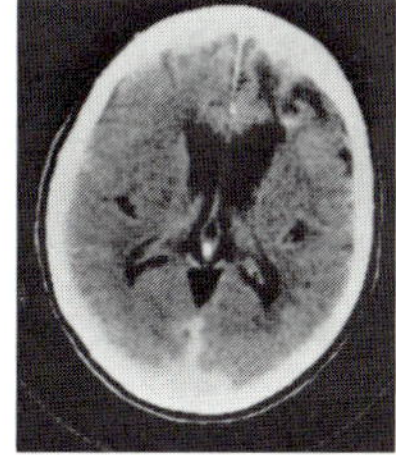

8.1. CT scan.

8.2. SPECT study in transaxial plane.

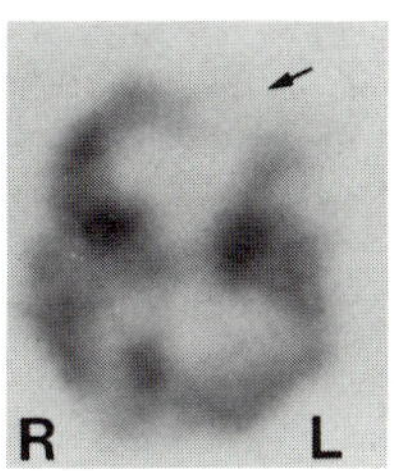

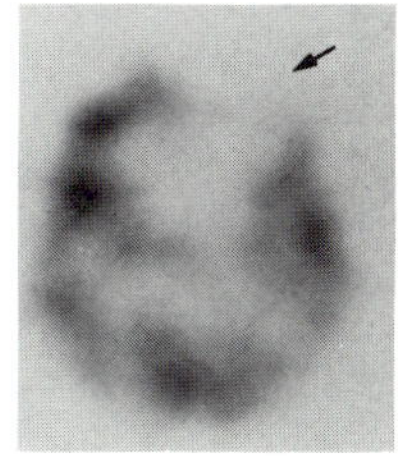

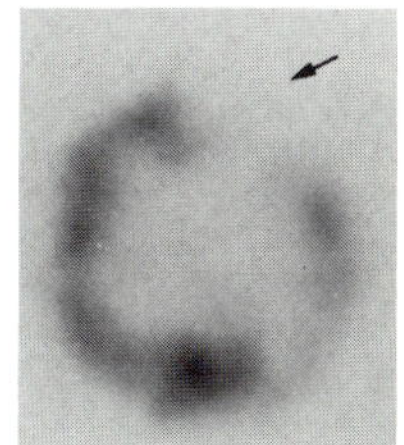

SECTION II: Cerebrovascular Disease

CASE 9: Contributed by Kazufumi Kimura, M.D., SPECT studies acquired on Hitachi Gamma View SPECT 2000 H-40 (Spectrum NeuroSpect 2000).

This patient had a known occlusion of the left internal carotid artery, with an infarction in the distribution of the left middle cerebral artery.

The cerebral SPECT study (Fig. 9.1) in the transaxial plane indicated an absence of tracer uptake in most of the left cerebral hemisphere. This tracer distribution pattern was consistent with an internal carotid artery occlusion.

9.1. SPECT study in transaxial plane.

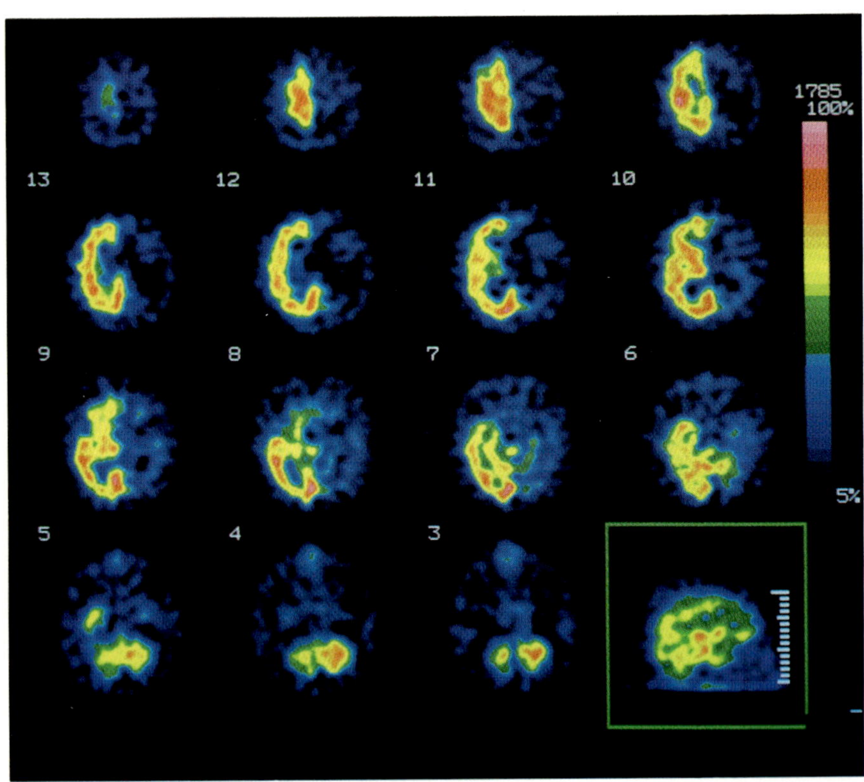

CASE 10: SPECT studies acquired on GE 400AC/T; STAR II.

One month following a stroke, this 80-year-old woman presented with progressive aphasia. A CT scan showed a left lacunar infarction and mild cortical atrophy.

A cerebral SPECT study (Fig. 10.1) in the transaxial (A), coronal (B), and sagittal (C) planes demonstrated an absence of tracer deposition in a large area of the left cerebral hemisphere, including the posterior frontal, parietal, temporal, and occipital lobes.

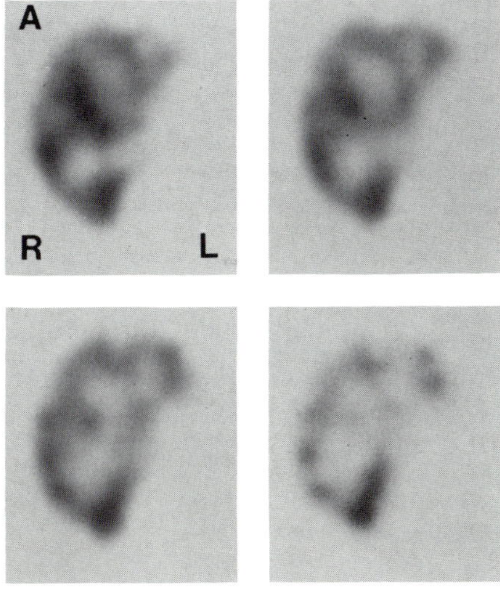

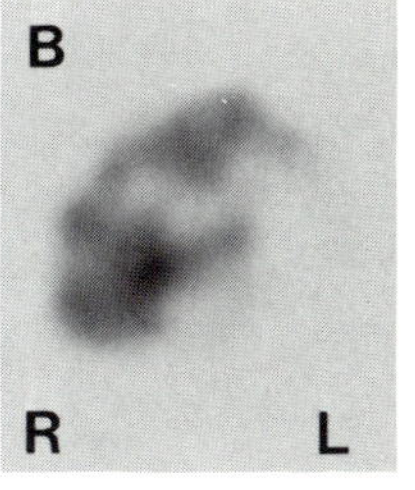

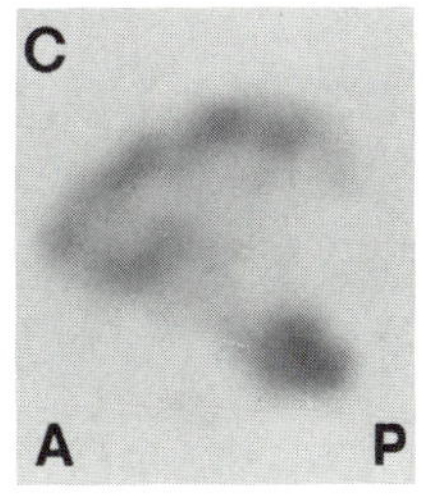

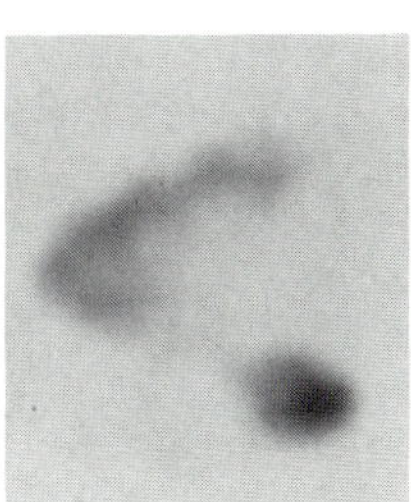

10.1. SPECT study. A: Transaxial; B: Coronal; and C: Sagittal planes.

CASE 11: SPECT studies acquired on GE 400AC/T; STAR.

Approximately two years after sustaining a left cerebrovascular accident, this 30-year-old, right-handed man was referred for evaluation. At the time of referral, the patient had a right hemiparesis, expressive aphasia, lingual dyspraxia, and a right homonymous hemianopsia.

A CT scan (Fig. 11.1) showed a large infarction of the left frontal, temporal, and anterior parietal lobes. A cerebral SPECT study (Fig. 11.2) in the transaxial (A), coronal (B), and sagittal (C) planes demonstrated absent tracer deposition in the region of the left middle cerebral artery. The extent of the abnormality appears greater on the SPECT study than on the CT.

Teaching Point:
In stroke patients where CT or MRI examinations do not explain the cause of the neurologic deficit, a cerebral SPECT study is an additional useful, noninvasive approach.

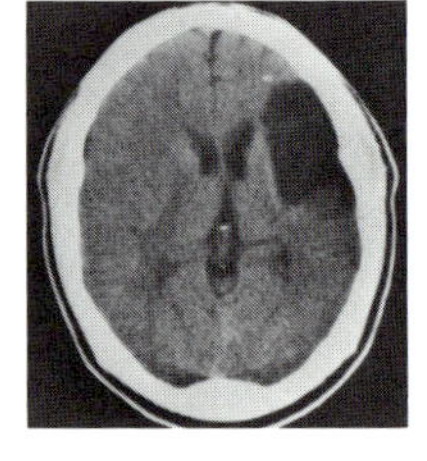
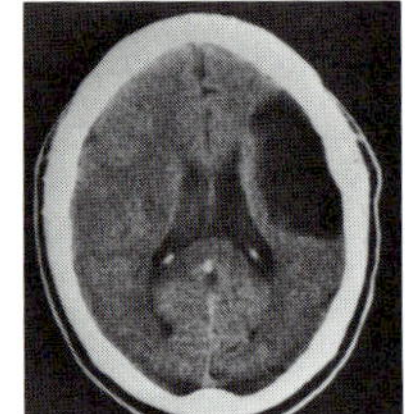

11.1. CT scan.

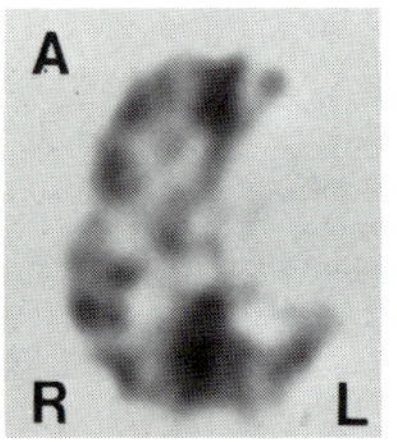

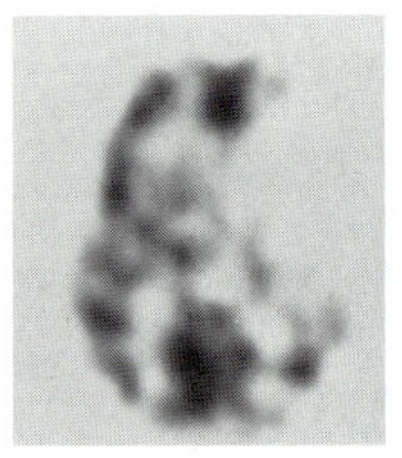
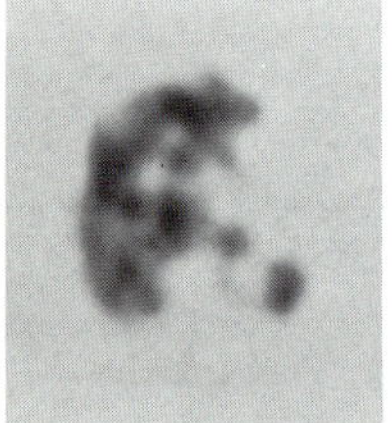
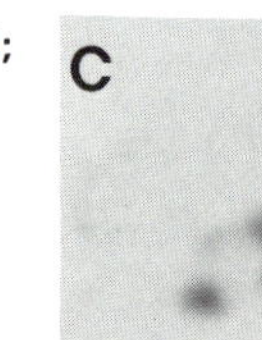

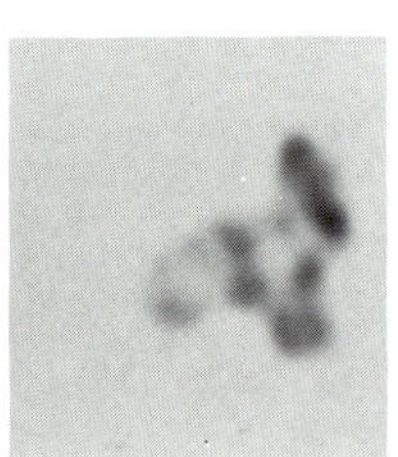

11.2. SPECT study. A: Transaxial; B: Coronal; and C: Sagittal planes.

CASE 12: SPECT studies acquired on GE 400AC/T; STAR.

This 45-year-old, right-handed woman had sustained a cerebrovascular accident of the left hemisphere two months before the cerebral SPECT study. At the time of the SPECT examination, she had a mild expressive/receptive aphasia and a moderate right hemiparesis in which the effect on the upper extremity was greater than that on the lower.

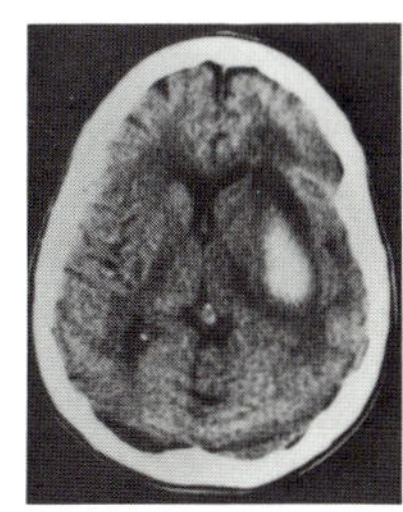

12.1. CT scan.

A CT Scan (Fig. 12.1) revealed a large hemorrhage in the region of the left basal ganglia. The cerebral SPECT study (Fig. 12.2) in the transaxial plane demonstrated an absence of tracer uptake at the site of the hemorrhage (arrow). In addition, a larger zone of decreased tracer deposition involved most of the left cerebral hemisphere (arrowheads) — a finding compatible with a larger area of associated ischemia.

12.2. SPECT study in transaxial plane.

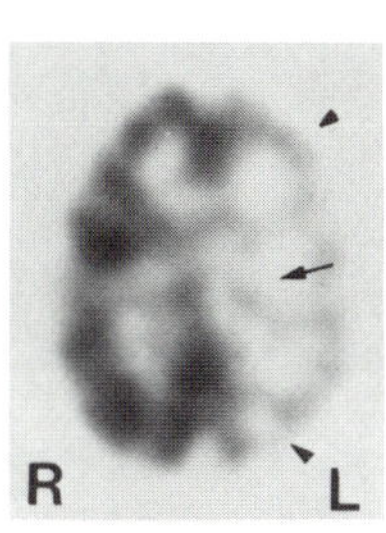

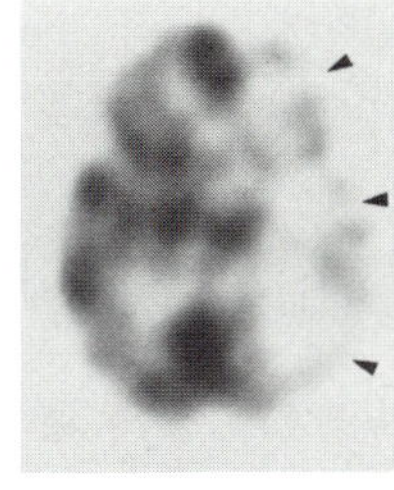

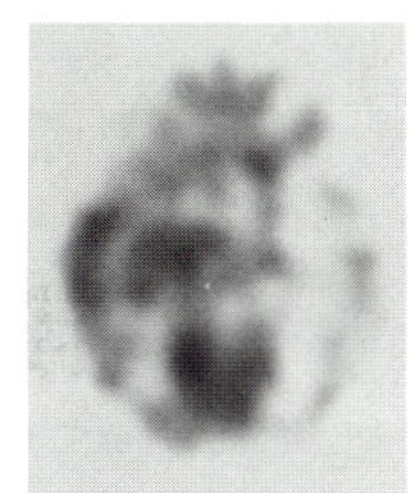

CASE 13: Contributed by Thomas C. Hill, M.D., SPECT studies acquired on Strichman SME-810.

This patient was referred for evaluation of an acute left cerebrovascular accident.

A CT scan (Fig. 13.1), performed without intravenous contrast, showed an oval low-density area involving the left basal ganglia and internal capsule.

A cerebral SPECT study (Fig. 13.2) in the transaxial plane showed that tracer deposition was absent in the left caudate nucleus (arrow). In addition, there was a slight decrease of tracer activity in the adjacent cortical gray matter (arrow head). The overall findings were consistent with a left lacunar infarction, with a concomitant decrease in the adjacent cerebral cortex.

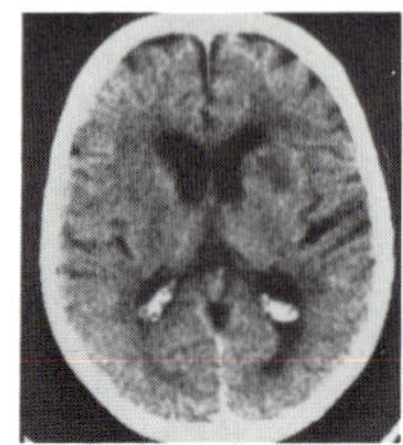

13.1. CT scan.

13.2. SPECT study in transaxial plane.

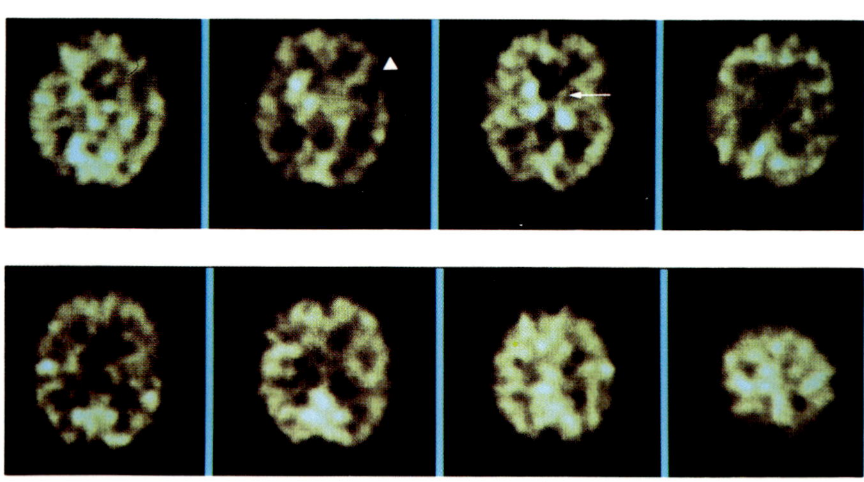

CASE 14: *SPECT studies acquired on GE 400AC/T; STAR II.*

This 62-year-old man was referred for evaluation of an acute onset of right-sided weakness, paresthesia, slurred speech, and decreased sensation on the right side of his face.

The CT scan (Fig. 14.1) showed a hypodensity in the region of the left caudate nucleus. A cerebral SPECT study (Fig. 14.2) in the transaxial plane revealed decreased tracer deposition in the left caudate nucleus (arrow). The combined findings are consistent with an acute left lacunar infarction.

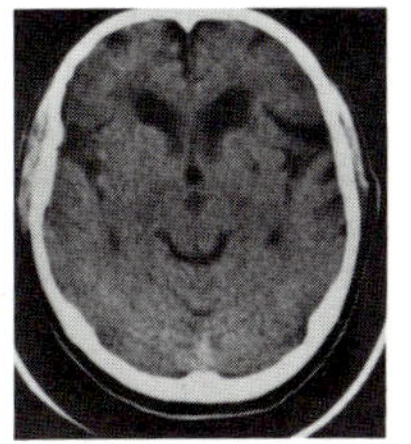

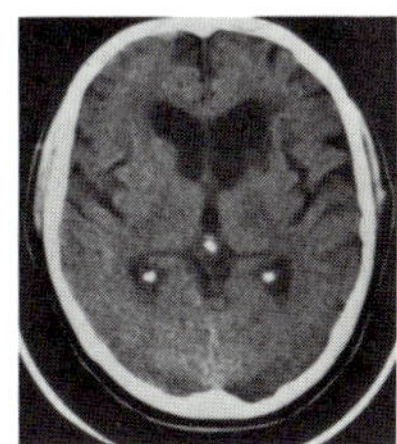

14.1. CT scan.

14.2. SPECT study in transaxial plane.

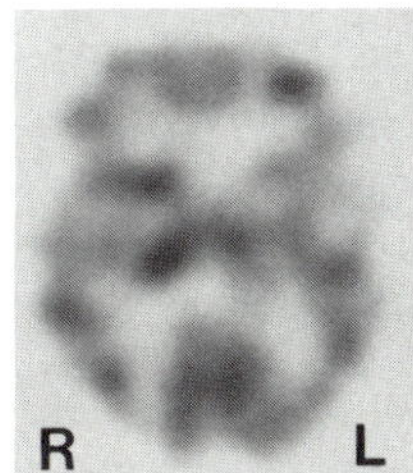

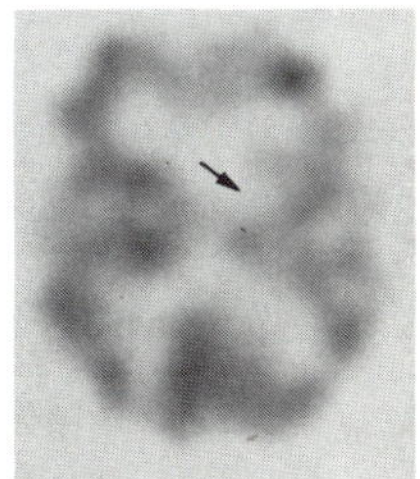

CASE 15: *Contributed by Robert L. Kagan, M.D., SPECT studies acquired on ADAC ARC-3000.*

This 60-year-old man was referred for evaluation of progressive loss of memory and increasing left-sided weakness and spasticity of 10 days' duration. He had previously (in 1975) sustained a right cerebrovascular accident with a left hemiparesis.

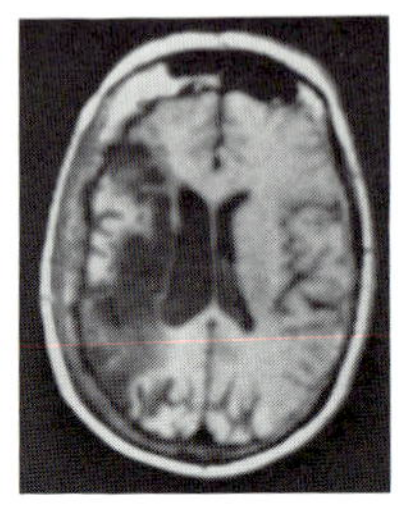

15.1. MRI study.

A MRI study (Fig. 15.1) demonstrated compensatory dilatation of the right lateral ventricle and a large area of encephalomalacia in the distribution of the right middle cerebral artery. The findings are consistent with an extensive infarction of the region of the right middle cerebral artery. A surrounding area of low signal-intensity may represent a zone of gliosis or ischemia around the region of the infarction.

An immediate cerebral SPECT study (Fig. 15.2 A) in the transaxial plane revealed absent tracer deposition (arrows); the delayed study (Fig. 15.2 B), following one hour of hyperbaric oxygen therapy, showed a smaller area of absent tracer deposition, which suggests an ischemic zone in the region surrounding the old infarction.

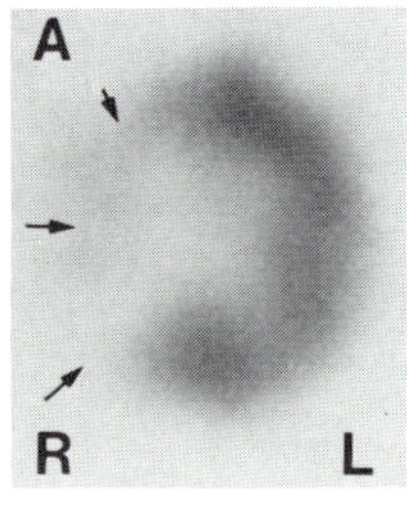

15.2A. Immediate SPECT study in transaxial plane.

Teaching Point:
The cerebral SPECT study may, at times, be useful for characterizing equivocal areas of abnormality noted on either MRI or CT scans.

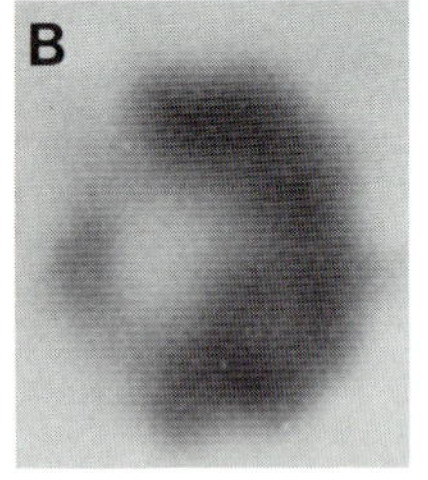

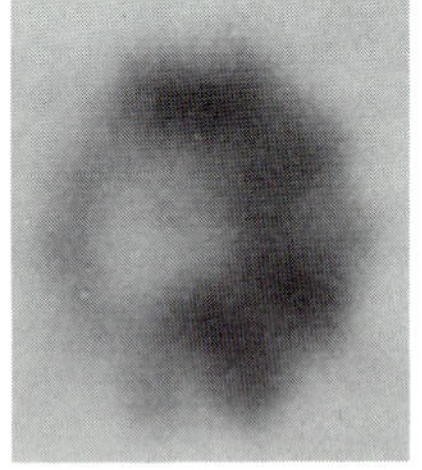

15.2B. Delayed SPECT study.

CASE 16: *SPECT studies acquired on GE 400AC/T; STAR.*

This 58-year-old, right-handed man was referred for evaluation of an old cerebrovascular accident of the right hemisphere. He had a long history of coronary artery disease and two years previously had sustained an infarction of the right middle cerebral artery. At evaluation, he showed a left hemiparesis and severe dysarthria.

A CT scan (Fig. 16.1) revealed an old right temporal and parietal infarction and mild dilatation of the right lateral ventricle.

A cerebral SPECT study (Fig. 16.2) in the transaxial (A), coronal (B), and sagittal (C) planes showed that tracer deposition was absent in the right posterior frontal, temporal, parietal, and occipital lobes. In addition, tracer deposition was reduced in the adjacent subcortical structures. The cerebral SPECT study showed a pattern that was more extensive than that seen on the CT scan.

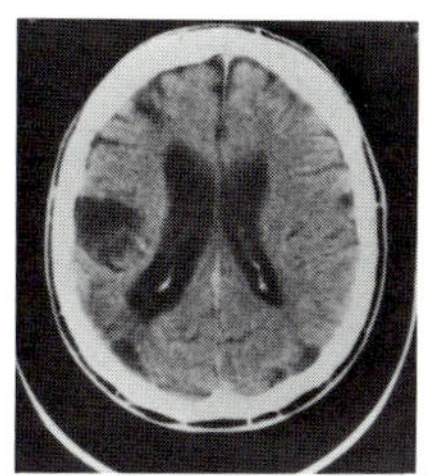

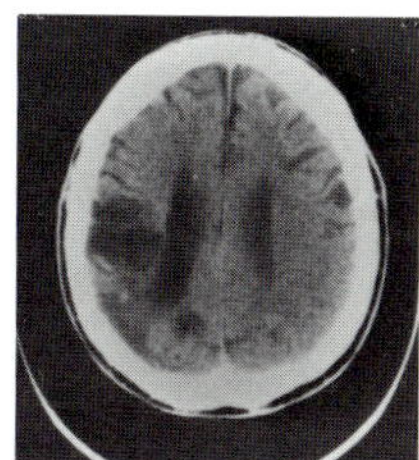

16.1. CT scan.

16.2. SPECT study. A: Transaxial; B: Coronal; and C: Sagittal planes.

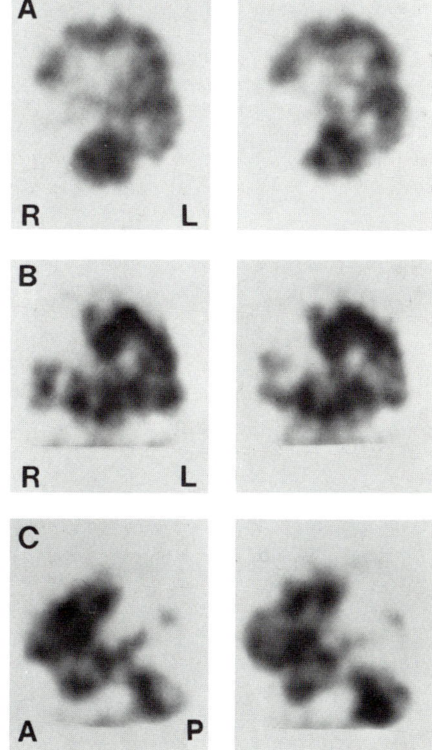

CASE 17: Contributed by Kazufumi Kimura, M.D., SPECT studies acquired on Hitachi Gamma View SPECT 2000H-40 (Spectrum NeuroSpect 2000).

This patient was referred for evaluation of a known intracerebral hemorrhage in the left occipital lobe.
The immediate cerebral SPECT study (Fig. 17.1A) in the transaxial and sagittal planes showed tracer deposition was absent in the left occipital lobe (arrows). The delayed study (Fig. 17.1B) showed a smaller area of absent deposition of tracer (arrow head), suggesting a surrounding zone of ischemia associated with the infarction.

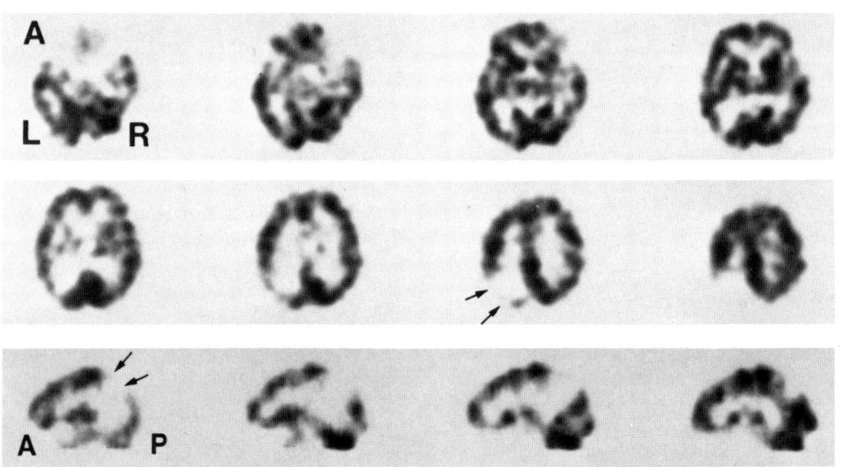

△17.1A. Immediate SPECT study in transaxial (upper/middle rows) and sagittal (lower row) planes.

▽17.1B. Delayed SPECT study.

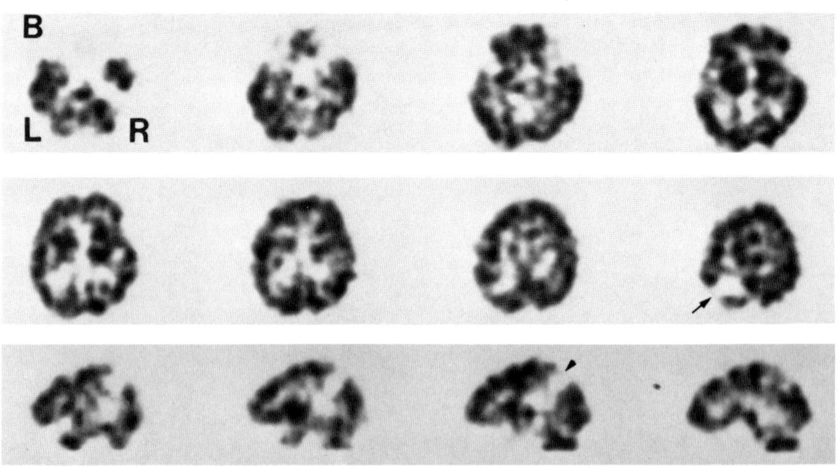

CASE 18: SPECT studies acquired on GE 400AC/T; STAR II.

This 60-year-old woman was referred for evaluation of a single episode of occipital headache, loss of vision, and transient loss of consciousness.

Serial CT scans (Fig. 18.1) showed an evolving infarction of the right occipital lobe, with involvement of the right temporal lobe and thalamus. In addition, an osteoma of the left parietal bone was noted to project from the inner table of the skull.

The cerebral SPECT study (Fig. 18.2) in the transaxial (A), coronal (B), and sagittal (C) planes showed an area of absent tracer deposition in the right occipital lobe, with involvement of the visual cortex (arrows). The adjacent posterior temporal lobe and posterior right thalamus were also found to be involved. The overall findings are quite typical of an infarction of the right posterior cerebral artery.

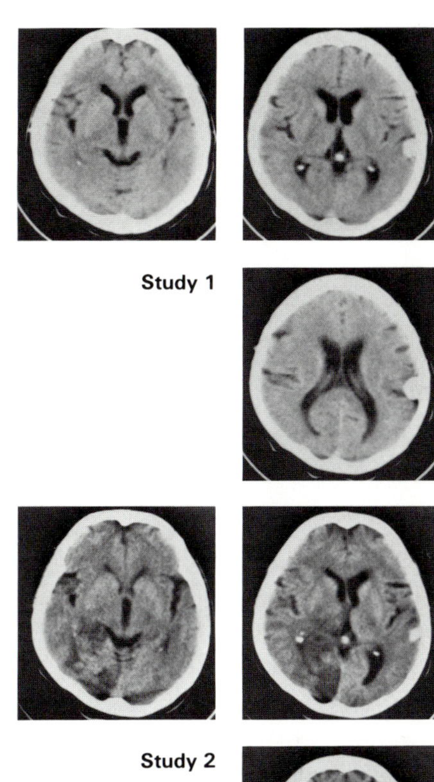

18.1. Serial CT scans.

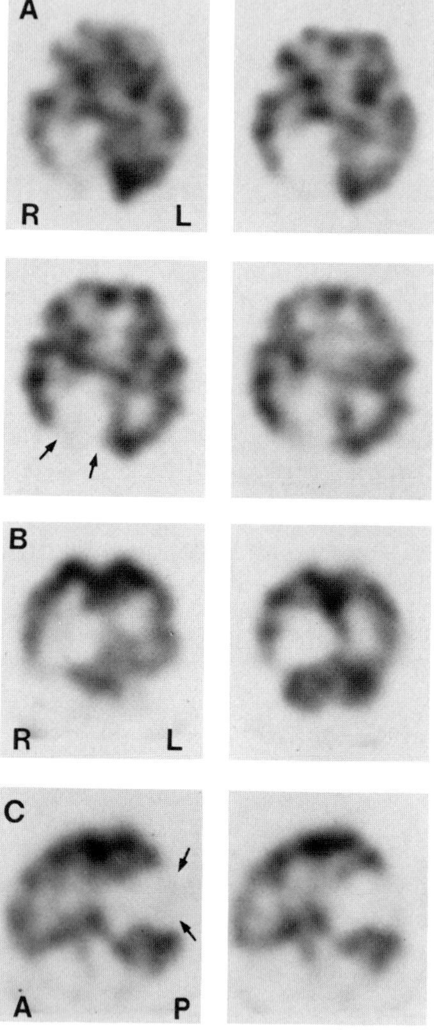

18.2. SPECT study. A: Transaxial; B: Coronal; and C: Sagittal planes.

__CASE 19:__ Contributed by Lee Katims, M.D., SPECT studies acquired on Toshiba-901 SPECT.

This 68-year-old man was referred for evaluation of progressive syncopal episodes. The patient had sustained a cerebrovascular accident approximately one year prior.

A CT scan (Fig. 19.1) revealed a small lacunar infarction within the left thalamus with a more extensive right posterior parietal-occipital region infarction. A smaller infarction in the left posterior parietal region was also noted.

The cerebral SPECT study (Fig. 19.2) in the transaxial (A), coronal (B), and sagittal (C) planes revealed diminished to absent tracer deposition in the right parietal-occipital region (arrow) and to a lesser degree in the left parietal lobe (arrowhead), corresponding to the areas of abnormality on the CT scan.

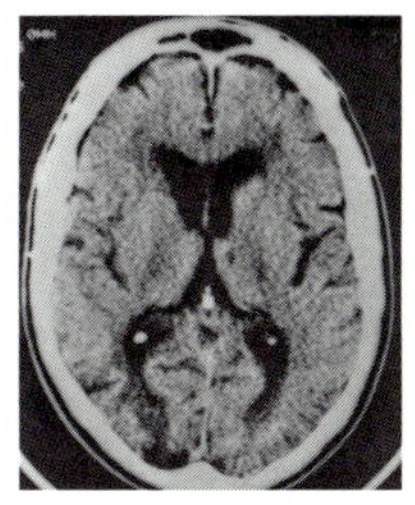

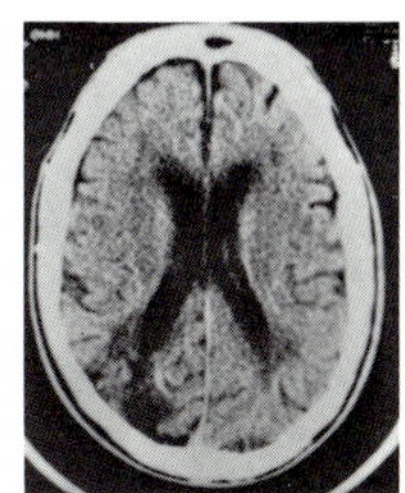

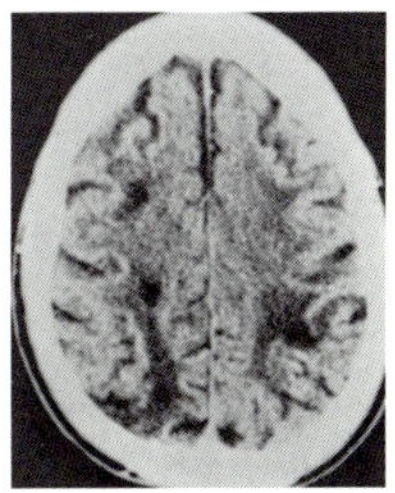

19.1. CT scan.

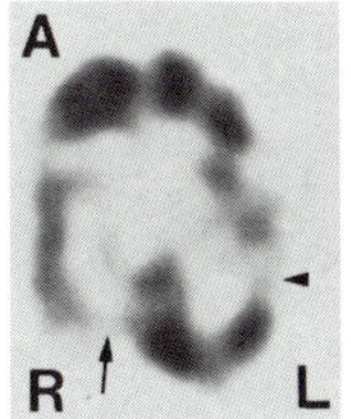

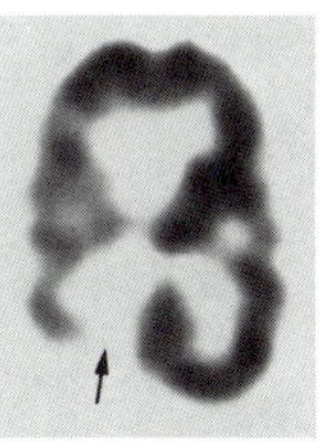

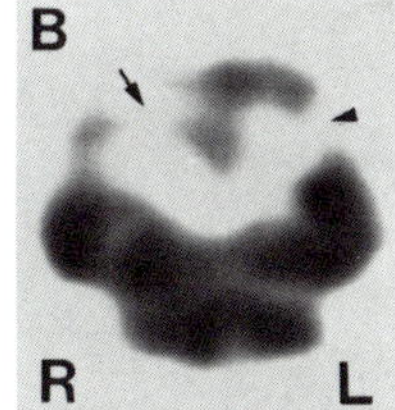

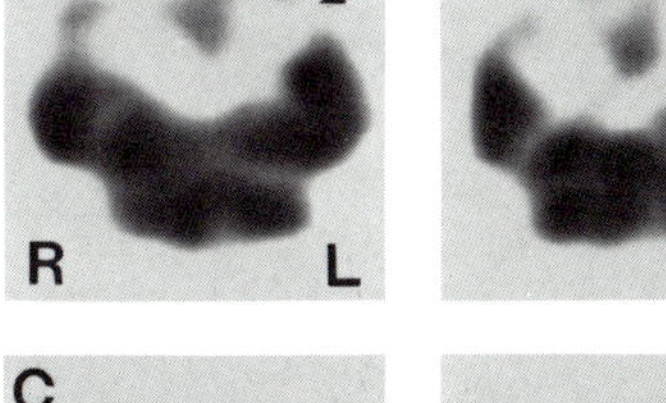

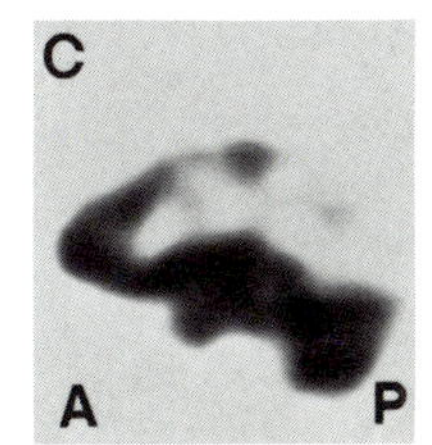

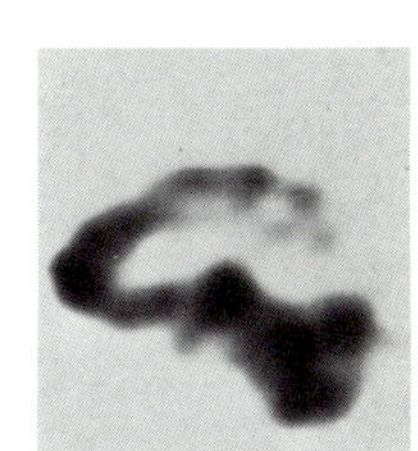

19.2. SPECT study. A: Transaxial; B: Coronal; and C: Sagittal planes.

CASE 20: Contributed by William L. Ashburn, M.D., SPECT studies acquired on Elscint APEX 409AG-ECT; APEX 009 Processor.

This 57-year-old man was referred for evaluation of dizziness of six months' duration.

A CT scan was negative. A cerebral SPECT study (Fig. 20.1) in the transaxial plane showed a marked decrease in tracer deposition in the left cerebellar hemisphere (arrows) which was compatible with ischemia or infarction in that region. A follow-up cerebral arteriogram revealed absent flow to the left cerebellar hemisphere due to complete occlusion of the left vertebral artery and 60% stenosis of both internal carotid arteries.

20.1. SPECT study in transaxial plane.

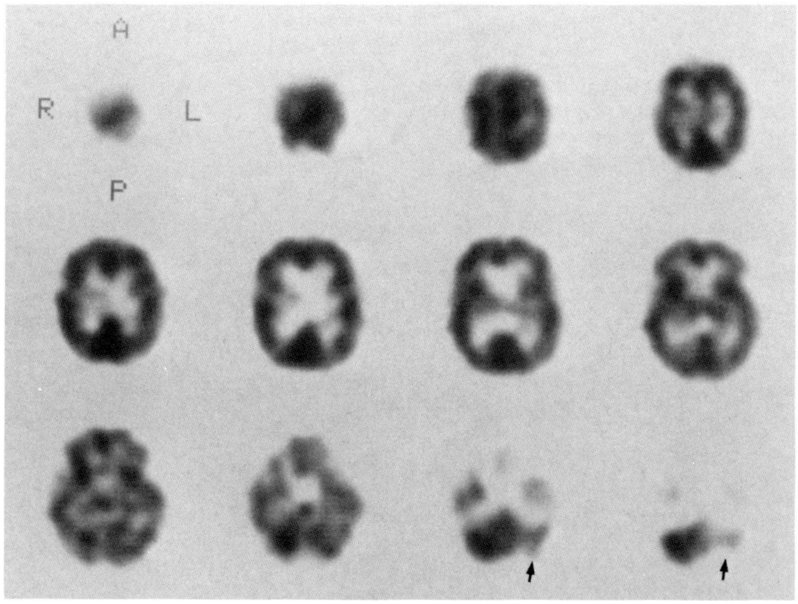

SECTION II: B

Dementias

INTRODUCTION

There has been a rapid increase in the utility of SPECT scanning in the diagnosis and understanding of dementia and its underlying, disordered brain function. PET imaging has demonstrated characteristic alterations in cerebral perfusion and metabolism in Alzheimer's disease, but its expense and lack of availability prevent it from being used routinely.

SPECT imaging is widely available and demonstrates the same patterns of perfusion and metabolic alteration in dementia as those seen with PET imaging. SPECT distinguishes Alzheimer's disease from other types of dementia, such as multi-infarct dementia, hydrocephalus, progressive supranuclear palsy, and various frontal lobe dementias. It also helps differentiate Alzheimer's disease from the cognitive impairments associated with depression.

The typical finding in Alzheimer's disease (Cases 1-3) is bilateral parietotemporal hypoperfusion and hypometabolism, with sparing of the sensorimotor and occipital regions. Frontal lobe hypoperfusion and hypometabolism are also found in more-advanced cases. In a smaller proportion of patients, the SPECT abnormalities are primarily unilateral, and they correspond to the signs of unilateral functional disturbances that are found on neuropsychological tests.

The second most-common type of dementia is multi-infarct dementia (MID). This dementia is characterized by a random, multifocal distribution of infarcts and abnormalities which are associated with specific vascular territories (Cases 4-6).

There are other important types of dementia (Cases 7-14): for example; the pseudodementia of depression, hydrocephalus; Huntington's disease; alcohol or other substance abuse; frontal lobe disease (such as Pick's); and, more recently, AIDS-related dementia. SPECT imaging is helpful in the differentiation of the pathophysiologic patterns seen in the various forms of dementia. It is also useful in establishing physiological subgroups that correlate with the different clinical presentations seen in patients with Alzheimer's and other diseases.

With PET imaging, the resolving power and availability of radiopharmaceuticals are superior; still, the availability, relative simplicity, and affordability of SPECT make it a valuable technique in the assessment of patients with dementia. It will play a significant role in the early diagnosis and longitudinal study of patients with dementia.

***CASE 1:** SPECT studies acquired on GE 400AC/T; STAR II.*

This 58-year-old woman presented with a two-year history of progressive deterioration of cognitive function.

A CT scan (Fig. 1.1) showed a moderate degree of cortical atrophy. The cerebral SPECT study (Fig. 1.2) in the transaxial (A), coronal (B), and sagittal (C) planes revealed bilateral decreased tracer deposition in the posterior parietal, temporal, and occipital lobes, and in the anterior frontal lobes. There was relatively normal deposition in the subcortical areas, the motor strip cortex, and cerebellum. These findings are typcial of the pattern seen with dementia of the Alzheimer's type.

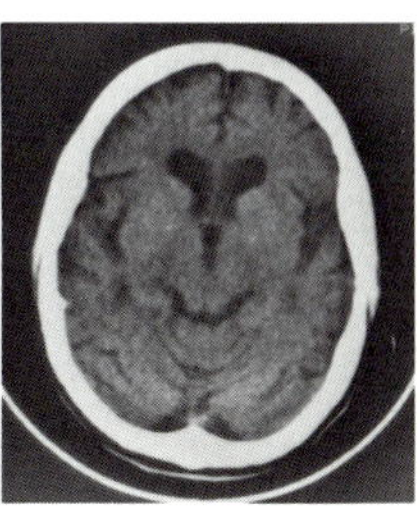

1.1. CT scan.

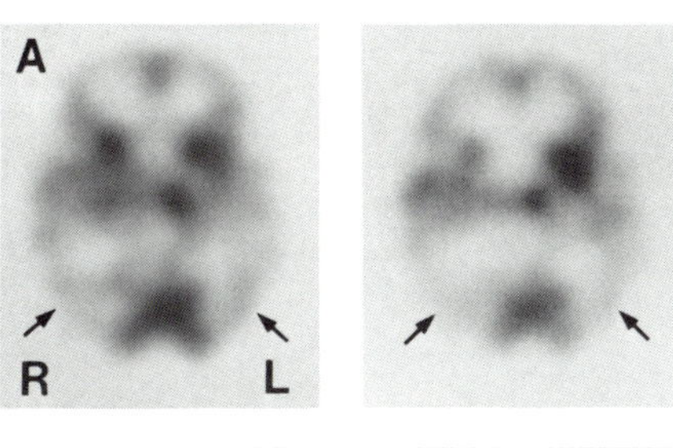

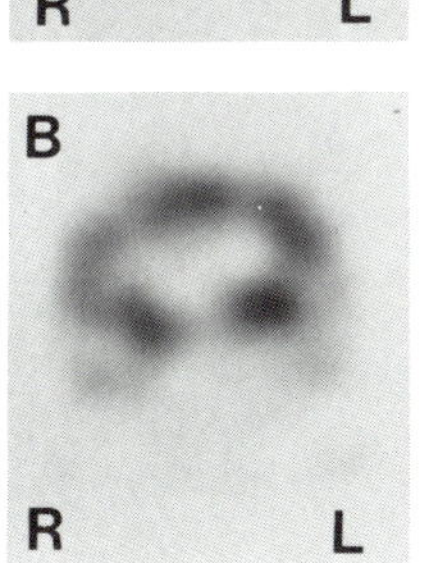

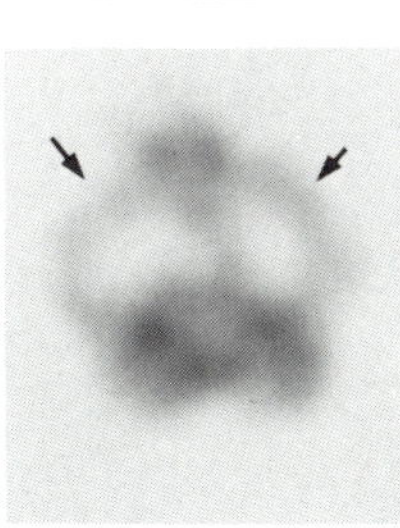

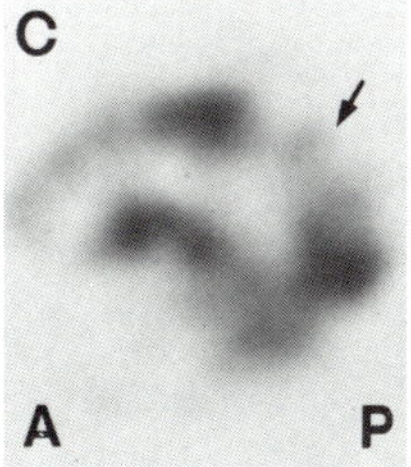

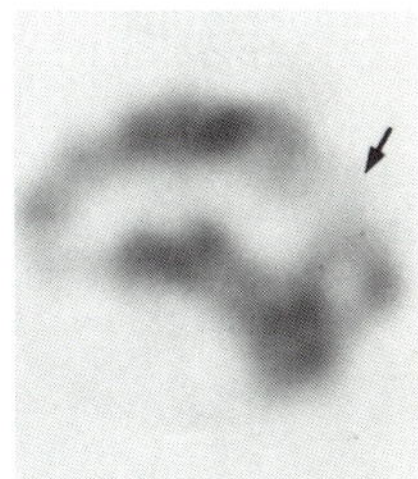

1.2. SPECT study. A: Transaxial; B: Coronal; and C: Sagittal planes.

CASE 2: *SPECT studies acquired on GE 400AC/T; STAR II.*

This 78-year-old man was referred for evaluation of a progressive dementia of six years' duration.

The CT scan (Fig. 2.1) revealed diffuse cortical atrophy. The cerebral SPECT study (Fig. 2.2) in the transaxial (A), coronal (B), and sagittal (C) planes revealed reduced tracer deposition throughout the cerebral cortex, with a relatively normal pattern in the cerebellum and subcortical regions. The decrease in cortical activity of the tracer was most pronounced in the posterior parietal-occipital areas (arrows), with relative sparing of the motor strip area. This pattern is most consistent with senile dementia of the Alzheimer's type.

Teaching Point:
The decreased tracer deposition in such areas as the posterior parietal region may be asymmetric, as demonstrated in this case.

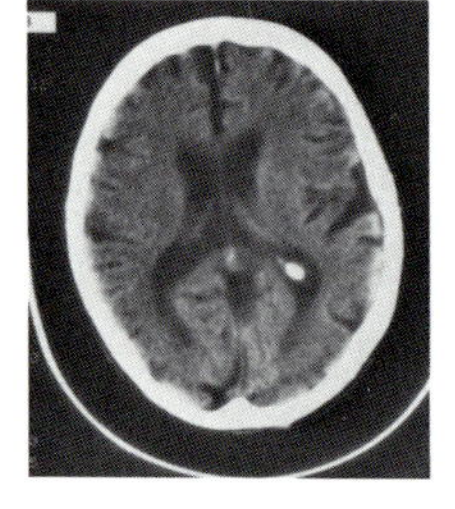
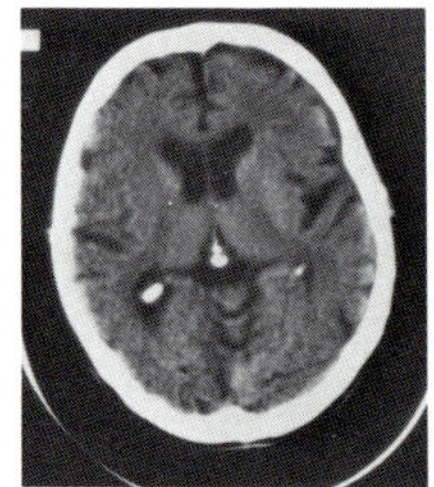

2.1. CT scan.

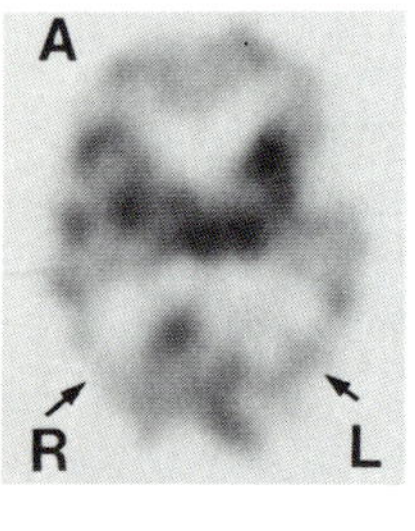

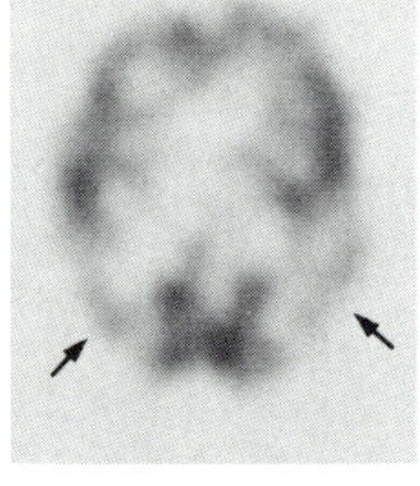
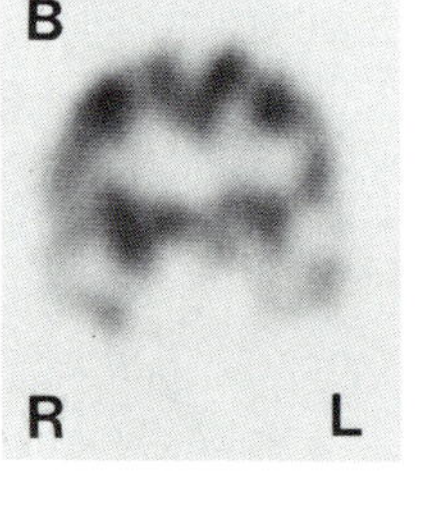

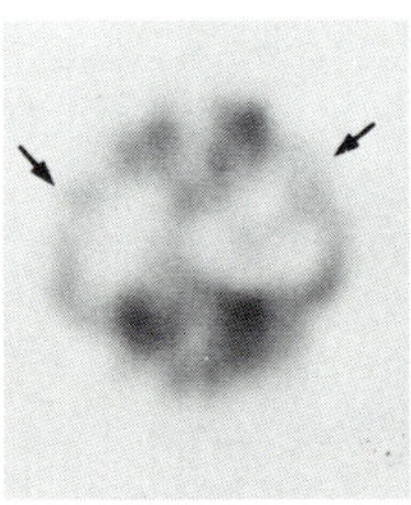
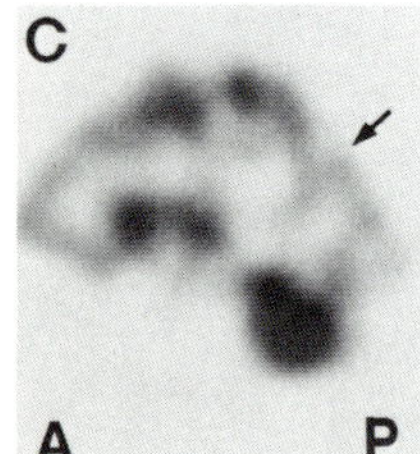

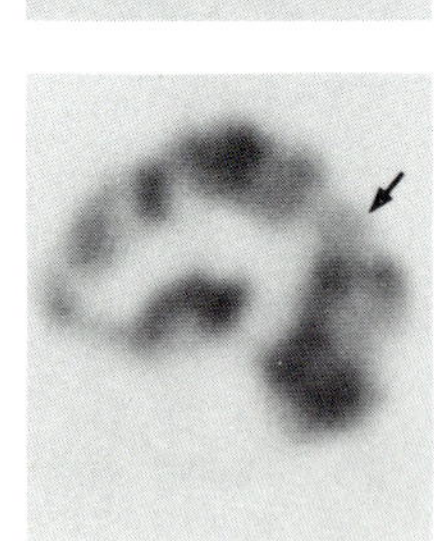

2.2. SPECT study. A: Transaxial; B: Coronal; and C: Sagittal planes.

CASE 3: *Contributed by Marvin B. Cohen, M.D., SPECT studies acquired on Siemens Rota Camera; (DEC) PDP 11/84 Computer.*

This 65-year-old man was referred for evaluation of probable Alzheimer's disease.

The cerebral SPECT study (Fig. 3.1) in the transaxial (A) and sagittal (B) planes revealed a mild bilateral decreased tracer deposition in the posterior parietal region (arrow). The pattern is consistent with the pattern seen in the early stages of Alzheimer's disease.

3.1. SPECT study. A: Transaxial and B: Sagittal planes.

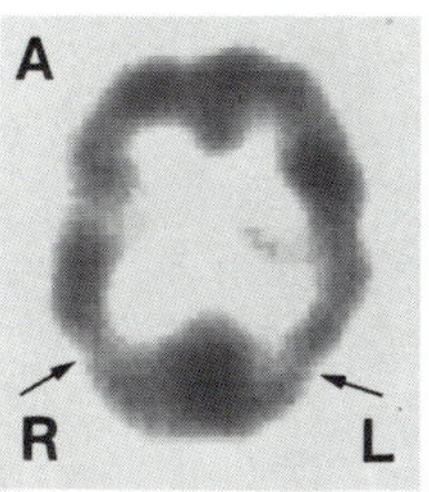

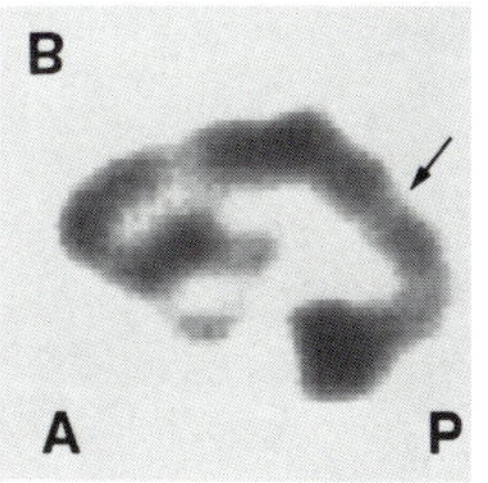

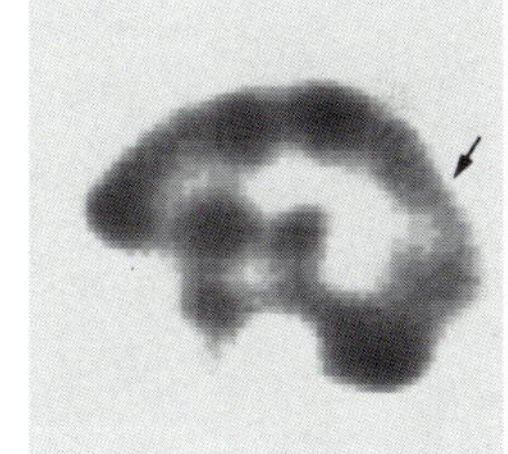

CASE 4: SPECT studies acquired on GE 400AC/T; STAR.

This 84-year-old, right-handed woman was referred for evaluation of a progressive deterioration of cognitive function.

The CT scan (Fig. 4.1) showed cerebral atrophy with dilatation of the lateral ventricles. The cerebral SPECT study (Fig. 4.2) in the transaxial plane revealed a marked decrease in tracer deposition throughout the periventricular white matter. In addition, a large area of diminished and absent deposition was noted in the left hemisphere, particularly in the posterior parietal-temporal and occipital lobes (arrows). The areas of involvement included the cortical territory surrounding the Sylvian fissure. Lesions in this area usually result in the significant alteration of language production and comprehension. This pattern is not typical of that usually associated with dementia of the Alzheimer's type.

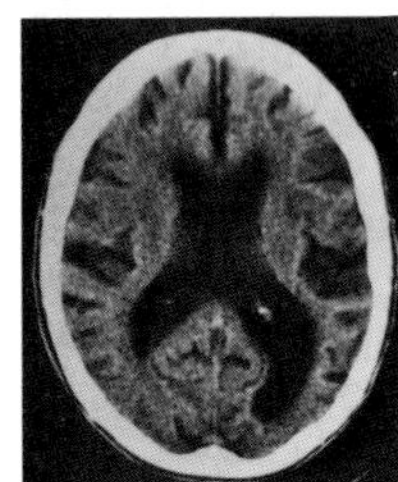

4.1. CT scan.

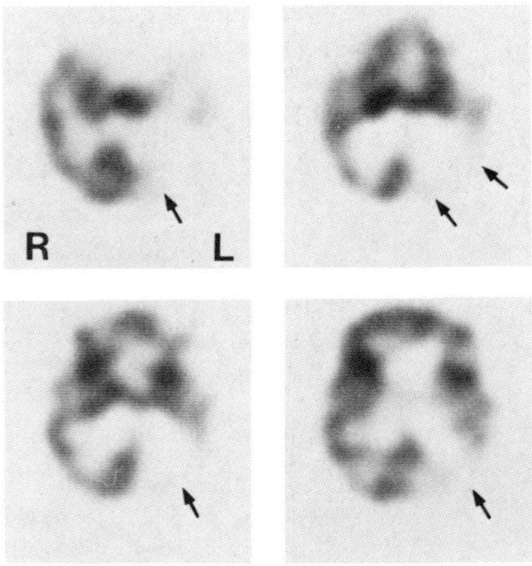

4.2. SPECT study in transaxial plane.

CASE 5: *SPECT studies acquired on GE 400AC/T; STAR II.*

This 64-year-old woman with known multi-infarct dementia was referred for evaluation of the recent onset of hemiplegia.

The CT scan (Fig. 5.1) demonstrated a large area of infarction involving the left occipital lobe. In addition, cortical atrophy and bilateral hypodensity of white matter, suggesting microvascular disease, were also noted.

A cerebral SPECT study (Fig. 5.2) in the transaxial (A), coronal (B), and sagittal (C) planes showed tracer deposition was absent in the left occipital lobe (arrows), left frontal lobe (arrow heads), and left basal ganglia.

Teaching Point:
In cases of multi-infarct dementia, the cerebral SPECT study will frequently be more useful than a CT scan alone for defining the full extent of disease.

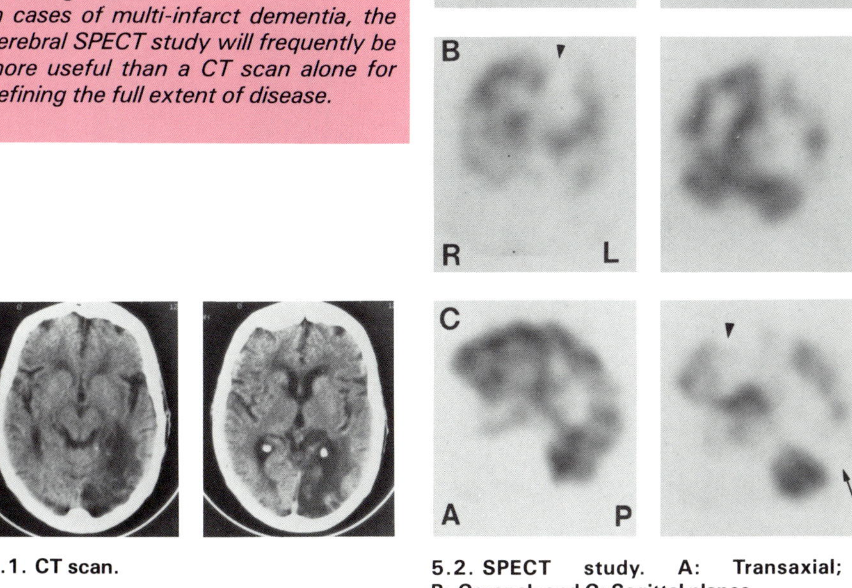

5.1. CT scan.

5.2. SPECT study. A: Transaxial; B: Coronal; and C: Sagittal planes.

CASE 6: *Contributed by Donald Fox, M.D. and Caesar Mayo, M.D., SPECT studies acquired on Spectrum 150 DT; SOPHA Computer.*

This 66-year-old woman was referred for evaluation of a progressive loss of memory over the preceding several months.

A MRI study (Fig. 6.1) showed a focal area of increased signal intensity, compatible with a lacunar infarction of indeterminate age. A cerebral SPECT study (Fig. 6.2) in the transaxial plane revealed tracer deposition was absent in the left occipital lobe (arrow head) and that tracer activity was decreased in the left basal ganglia (arrow).

Teaching Point:
The pattern in this case is compatible with a vascular etiology for the patient's dementia, rather than Alzheimer's disease.

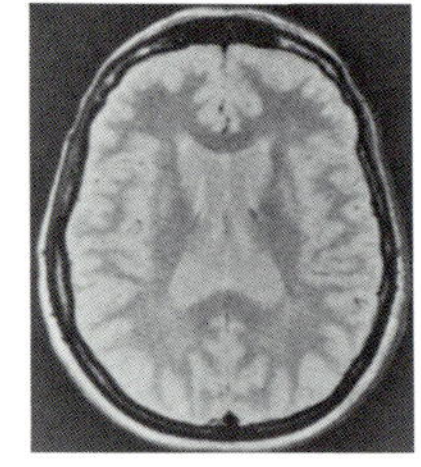

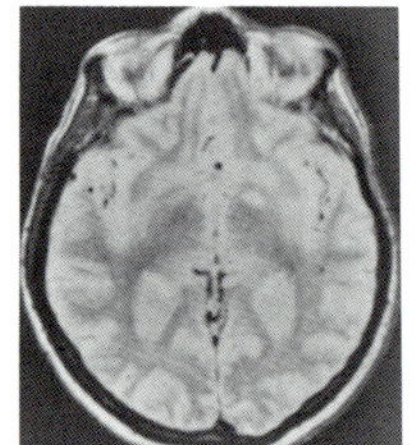

6.1. MRI study.

6.2. SPECT study in transaxial plane.

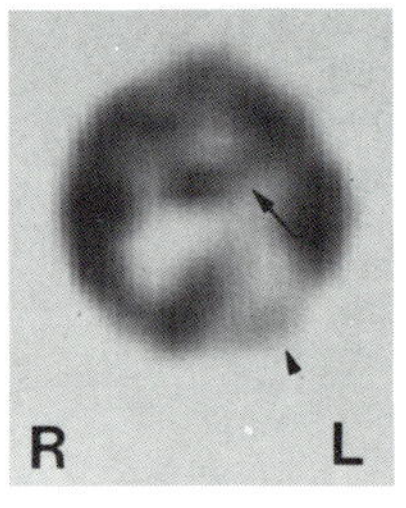

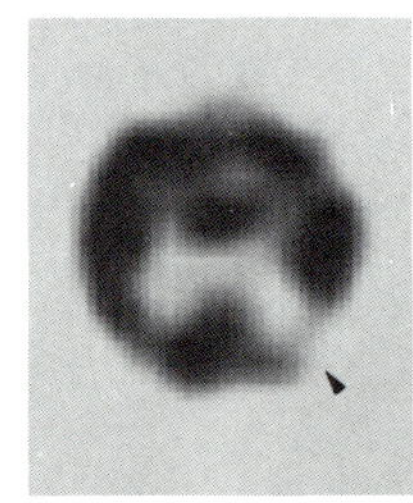

CASE 7: *SPECT studies acquired on GE 400AC/T; STAR II.*

This 77-year-old man presented with mild dementia and severe depression. An EEG revealed diffuse slow wave activity. The CT scan (Fig. 7.1) revealed moderate cortical atrophy.

A cerebral SPECT study (Fig. 7.2) in the transaxial (A), coronal (B), and sagittal (C) planes revealed reduced tracer deposition in the frontal and anterior parietal lobes (arrows). This pattern was more pronounced on the left side. In addition there was a mild degree of decreased tracer activity in the periventricular white matter corresponding to the cortical atrophy noted on the CT scan.

Teaching Point:
The overall findings on the cerebral SPECT study are consistent with a pattern of severe depression as a cause of the patient's symptoms.

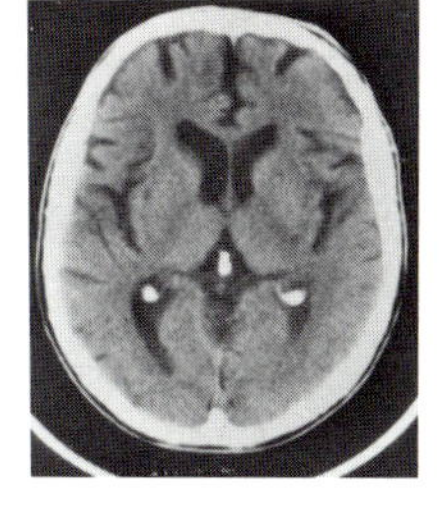
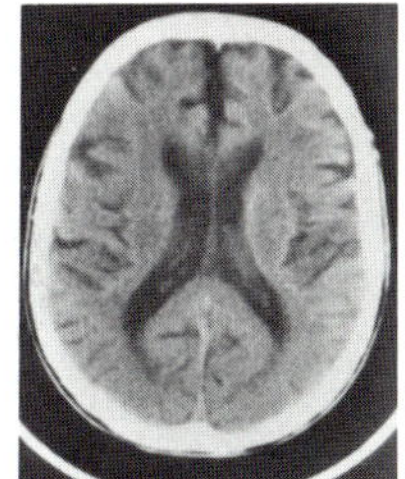

7.1. CT scan.

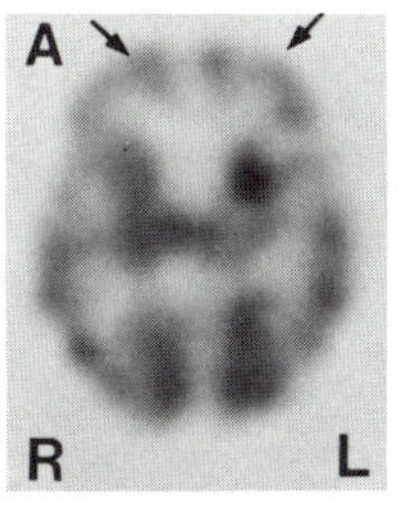

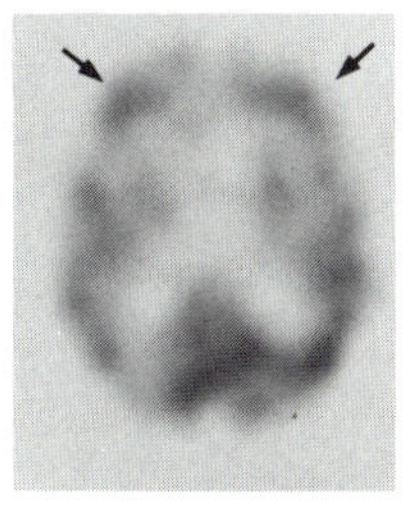
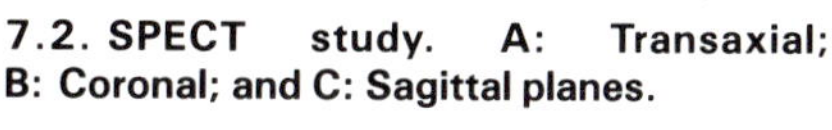

7.2. SPECT study. A: Transaxial; B: Coronal; and C: Sagittal planes.

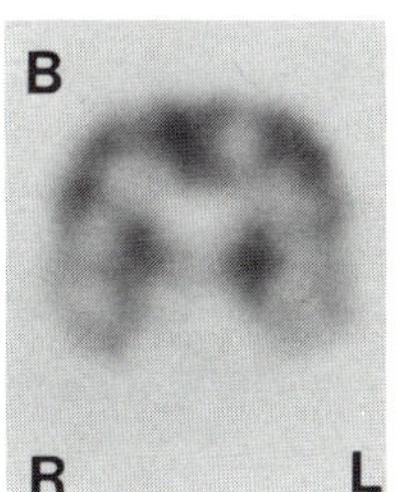

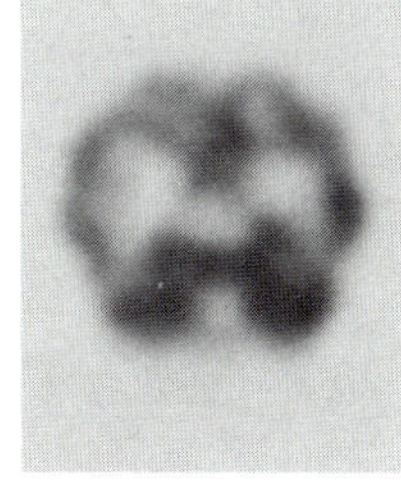
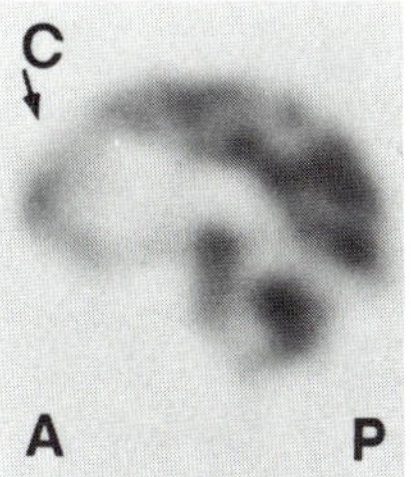

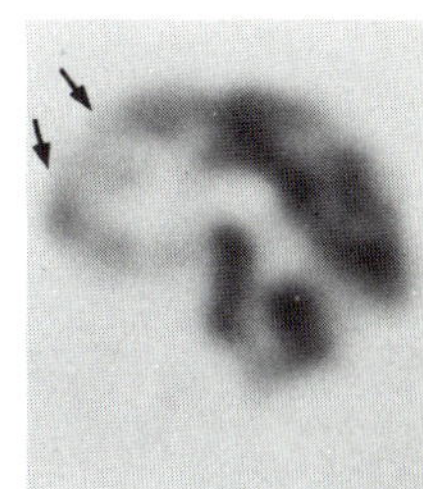

CASE 8: *SPECT studies acquired on GE 400AC/T; STAR.*

This 55-year-old woman was referred for evaluation of a rapid onset of confusion, ataxia, and urinary incontinence. An initial MRI study (Fig. 8.1) showed markedly enlarged lateral and third ventricles and abnormal periventricular high signal.

A cerebral SPECT study (Fig. 8.2) in the transaxial plane showed that tracer activity was reduced in the periventricular white matter, corresponding to the ventricular dilatation demonstrated by the MRI scan. In addition, tracer activity was decreased in the right occipital lobe.

Two months after a ventriculoperitoneal shunt procedure, a repeat MRI study (Fig. 8.3) demonstrated a significant decrease in the size of the lateral ventricles, with a VP shunt in place. A follow-up cerebral SPECT study (Fig. 8.4), done the following day, showed a normal, symmetric distribution of tracer. Clinically, the patient appeared normal in all physical and mental functions.

Teaching Point:
The pattern in this case is typical of the findings noted in normal pressure hydrocephalus.

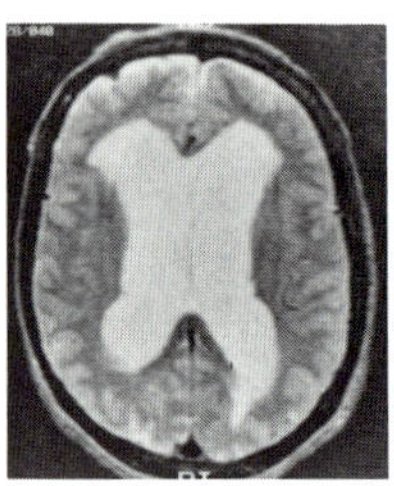

8.1. Initial MRI study.

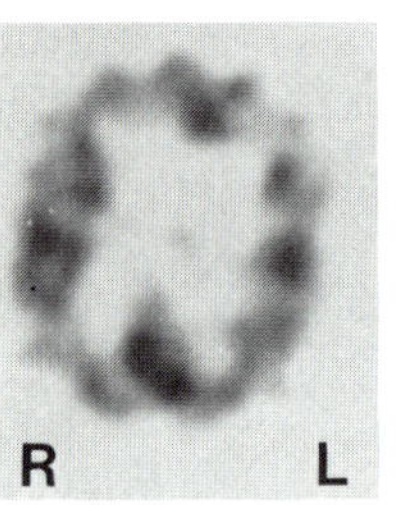

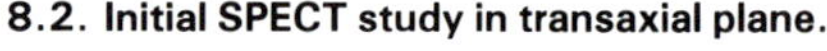

8.2. Initial SPECT study in transaxial plane.

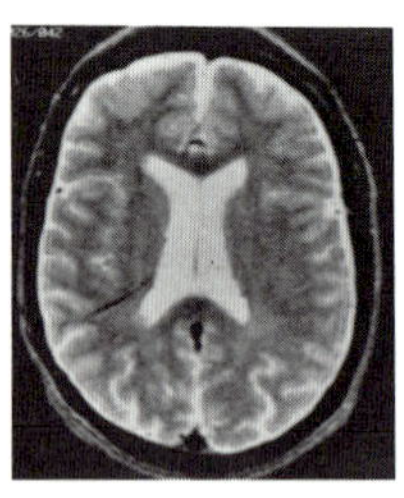

8.3. Follow-up MRI study.

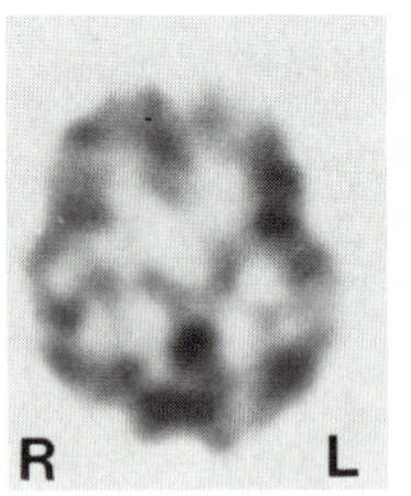

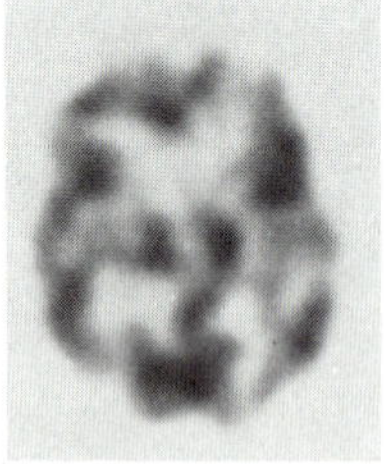

8.4. Follow-up SPECT study.

CASE 9: SPECT studies acquired on GE 400AC/T; STAR II.

This 55-year-old man was referred for an ongoing evaluation of his known Huntington's disease with mild dementia of nine years' duration. A CT scan (Fig. 9.1) showed diffuse atrophy of the cerebral cortex. The cerebral SPECT study (Fig. 9.2) in the transaxial (A), coronal (B), and sagittal (C) planes revealed decreased tracer deposition throughout the cerebral cortex and periventricular white matter, corresponding to the cortical atrophy noted on the CT scan. In addition, tracer deposition was absent bilaterally in the caudate nuclei (arrows).

Teaching Point:
Absent or markedly diminished tracer activity in the caudate nuclei is a frequent and characteristic pattern observed in patients with Huntington's disease.

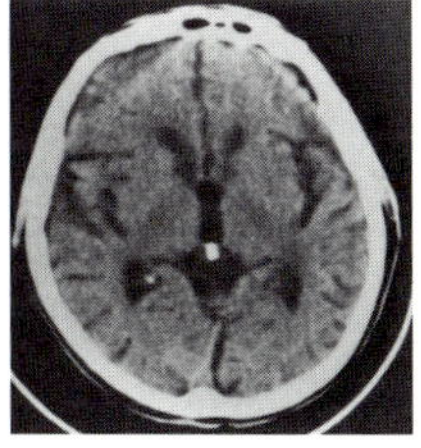

9.1. CT scan.

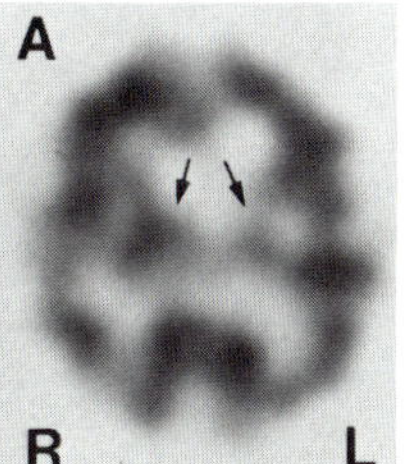

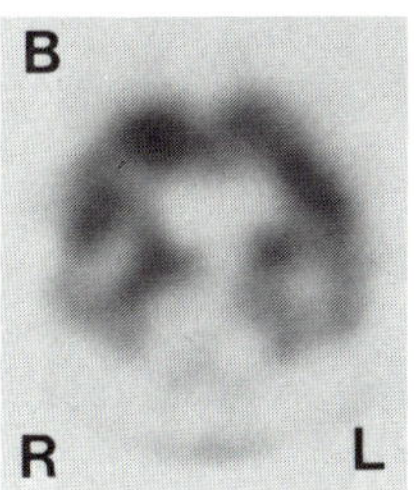

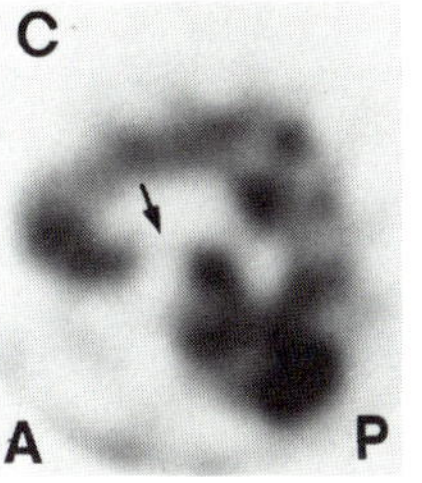

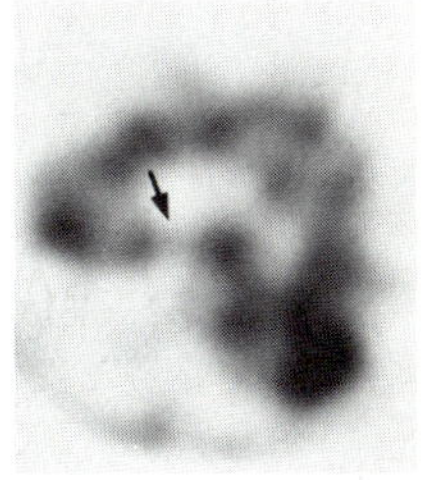

9.2. SPECT study. A: Transaxial; B: Coronal; and C: Sagittal planes.

CASE 10: *SPECT studies acquired on GE 400AC/T; STAR II*

This 45-year-old male was referred for evaluation of progressive dementia and athetosis. The patient's mother had previously been given a diagnosis of Huntington's disease.

A cerebral SPECT study (Fig. 10.1) in the transaxial (A), coronal (B), and sagittal (C) planes revealed a marked decrease in tracer deposition in the caudate nuclei (arrows). The findings were felt to be compatible with Huntington's disease.

Teaching Point:
This case presents an earlier form of Huntington's disease. In this particular case, the cerebral SPECT study was helpful in establishing the final diagnosis.

10.1. SPECT study. A: Transaxial; B: Coronal; and C: Sagittal planes.

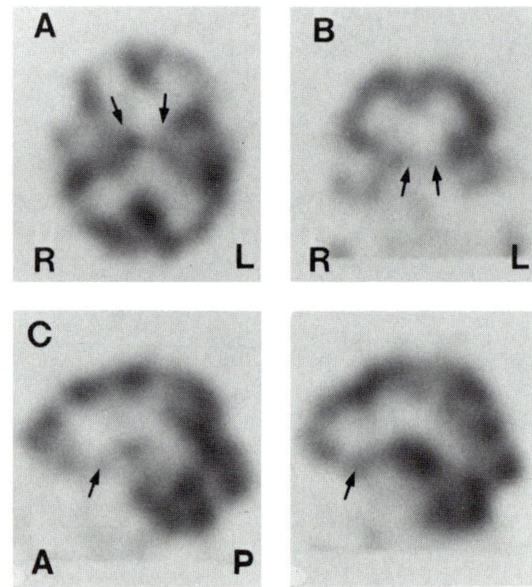

CASE 11: SPECT studies acquired on GE 400AC/T; STAR II.

This 26-year-old man was referred for evaluation of a progressive loss of memory. He had a long history of alcohol abuse, with fequent episodes of violent behavior when intoxicated.

A CT scan (Fig. 11.1) was within normal limits. A cerebral SPECT study (Fig. 11.2) in the transaxial (A), coronal (B), and sagittal (C) planes showed a very heterogeneous pattern of tracer distribution throughout the cerebral cortex and cerebellar hemispheres.

Teaching Point:
This pattern of tracer distribution is typical of alcohol and drug abuse patients. This heterogeneous distribution is quite different from the patterns seen in Alzheimer's disease, vascular dementia, and pseudodementia of major depression.

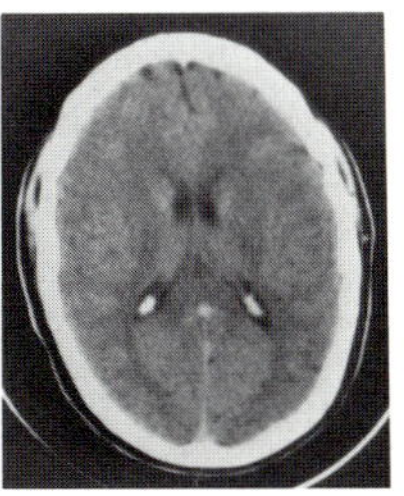

11.1. CT scan.

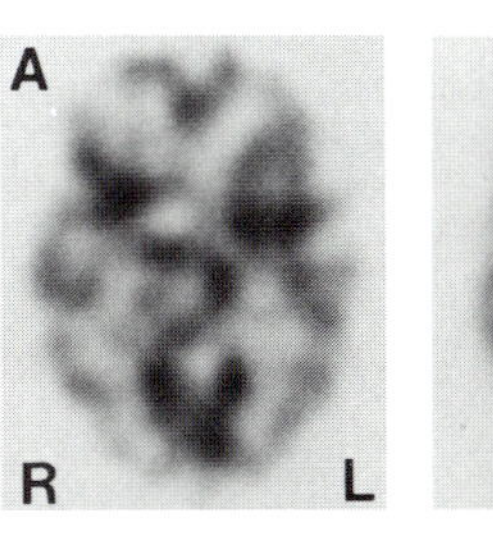

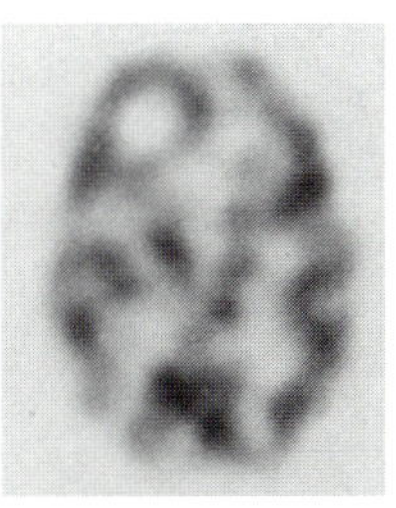

11.2. SPECT study. A: Transaxial; B: Coronal; and C: Sagittal planes.

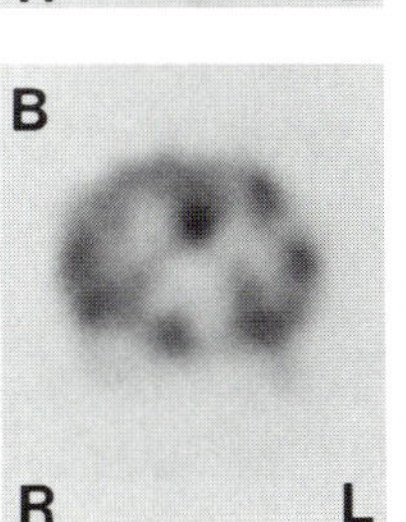

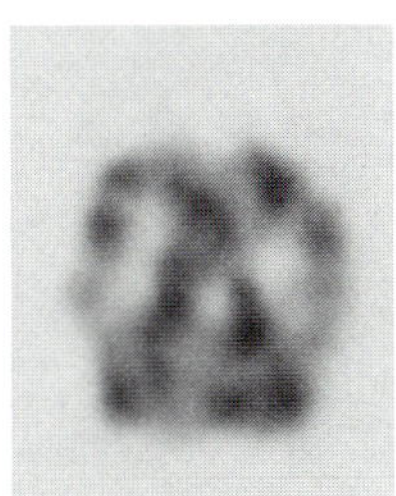

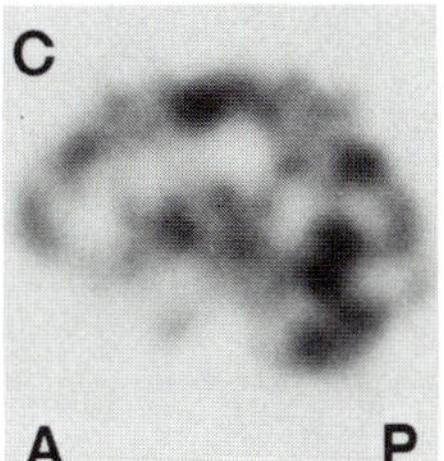

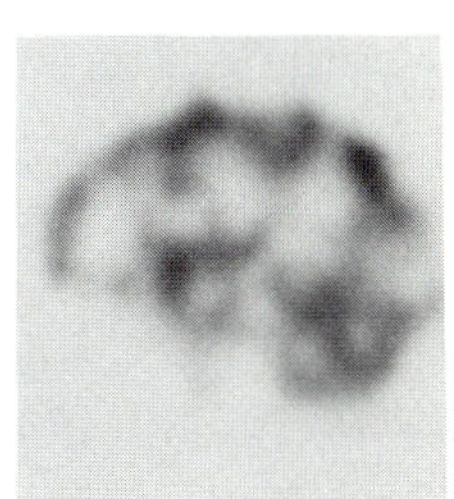

CASE 12: SPECT studies acquired on GE 400AC/T; STAR II.

This 37-year-old bisexual man was referred for evaluation of acute onset dementia; he also complained of blurred vision and generalized headaches. His neurological examination revealed a right hemiparesis and an expressive aphasia. Laboratory tests showed him to be seropositive for HIV antigen.

A CT scan (Fig. 12.1) was negative. A SPECT study (Fig.12.2) in the transaxial (A), coronal (B), and sagittal (C) planes showed decreased tracer deposition in the left frontal and parietal-occipital lobes (arrowheads), with a slight decrease in tracer activity in the right parietal-occipital lobe (open arrow). Tracer deposition was decreased in the thalamus, particularly on the left side (arrow).

Teaching Point:
As shown in this case, the cerebral SPECT study may be positive despite a negative CT or MRI scan. The typical pattern in AIDS related dementia shows random cortical defects combined with subcortical defects.

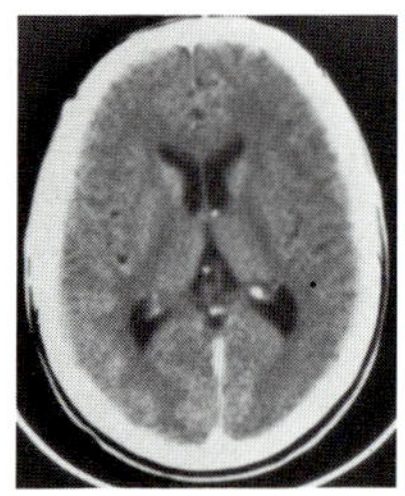

12.1. CT scan.

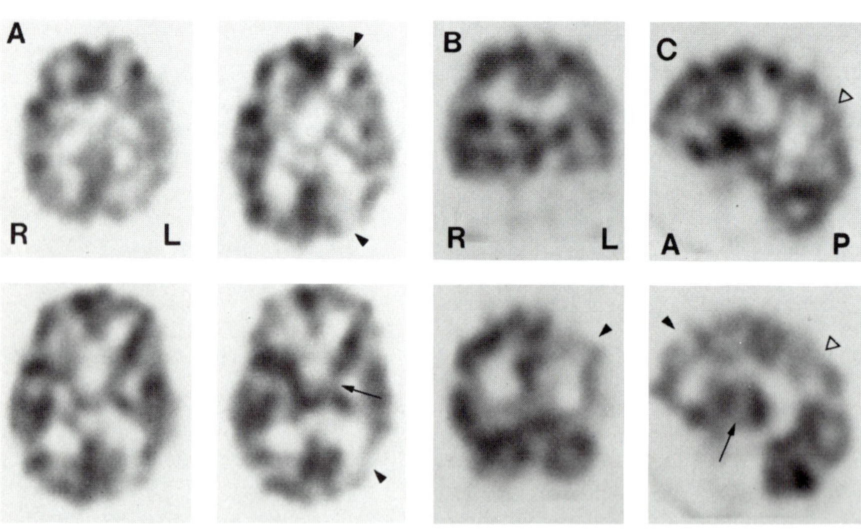

12.2. SPECT study. A: Transaxial; B: Coronal; and C: Sagittal planes.

CASE 13: *Contributed by David L. Kaufman, M.D., SPECT studies acquired on GE 400AC/T; STAR II.*

This 40-year-old homosexual man, known to be seropositive for HIV, was referred for evaluation of a progressive loss of concentration over three months.

A cerebral SPECT study (Fig. 13.1) in the transaxial (A), coronal (B), and sagittal (C) planes showed absent tracer deposition in the left frontal lobe laterally. The area of absent activity appeared to be extracerebral. As a consequence of this finding, a follow-up CT scan (Fig. 13.2) was performed; it demonstrated a large cystic area at the lateral aspect of the left frontal lobe.

Teaching Point:
The findings on cerebral SPECT studies may sometimes be nonspecific. In such instances, CT or MRI scans should be correlated with the SPECT studies to better characterize the abnormality.

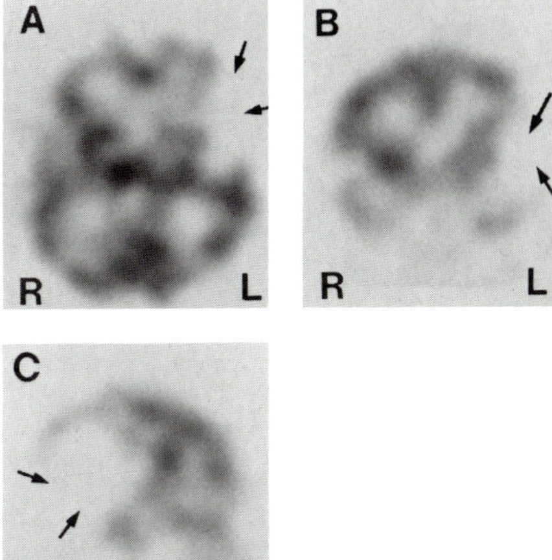

13.1. SPECT study. A: Transaxial; B: Coronal; and C: Sagittal planes.

13.2. Follow-up CT scan.

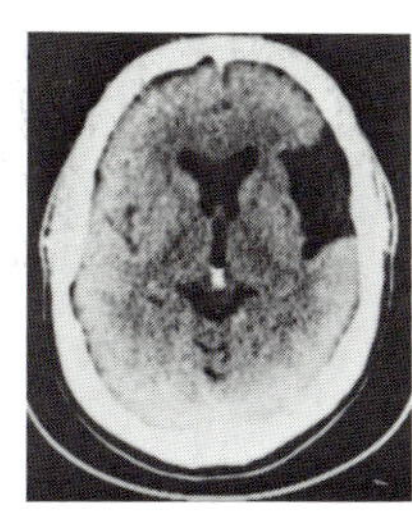

CASE 14: *SPECT studies acquired on GE 400AC/T; STAR II.*

This 39-year-old homosexual man was referred for evaluation of severe headaches of one month's duration. In addition, he reported blurred vision, intermittent chills, and night sweats. Neurological examination revealed a left homonymous hemianopsia.

A CT scan showed a large right parietal-occipital hypodensity with ring enhancement, with a mass effect on the surrounding structures. A cerebral SPECT study (Fig. 14.1) in the transaxial plane demonstrated absent tracer deposition in the right parietal-occipital and posterior temporal areas (arrows), corresponding to the area of abnormality on the CT scan. The final diagnosis was cerebral toxoplasmosis.

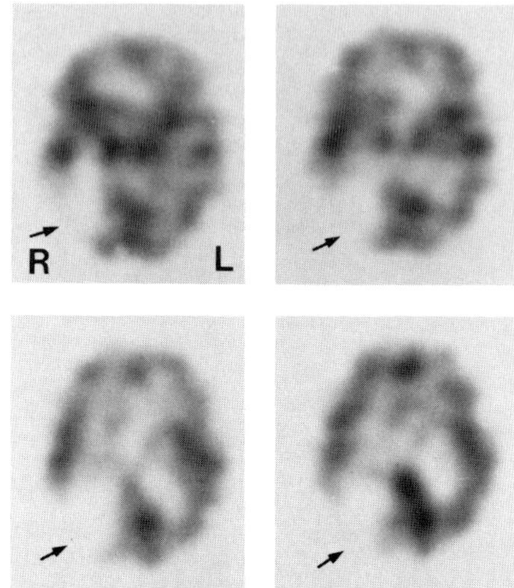

14.1. SPECT study in transaxial plane.

SECTION II: C

Seizure Disorders

INTRODUCTION

The clinical application of SPECT imaging in the study of patients with complicated epilepsy has rapidly expanded. SPECT imaging has gained a major role in the identification of epileptogenic foci in patients with medically intractible complex seizures who are being considered for surgery. In one respect, SPECT may be superior to PET: while PET requires that both the administration of an appropriate radiopharmaceutical and actual imaging take place during seizure activity, SPECT requires only that the radiopharmaceutical be injected during the seizure and permits imaging of cerebral perfusion and metabolic changes up to four or five hours after seizure activity has ceased. This allows sufficient time for patients to be brought to an imaging area, and it increases the probability of identifying the physiologic changes associated with an active seizure focus. Seizure patterns observed on SPECT imaging parallel those reported with PET. During the interictal period, in approximately 50% of patients, the seizure focus is identified as an area of reduced perfusion and metabolism; during an active seizure in approximately 80% of patients, an area of increased perfusion and metabolism mark the active seizure focus. Therefore, SPECT imaging, with its wide clinical availability, is becoming a key method for identifying single and multiple seizure foci, and minimizing the need for invasive electrophysiological studies. Used in conjunction with CT, MRI, and electroencephlographic studies, SPECT increases the accuracy of lesion identification and improves the rate of surgical success.

CASE 1: SPECT studies acquired on GE 400AC/T; STAR II.

This 31-year-old man was referred for evaluation of a recent seizure with associated occipital headaches, photophobia, and intermittent flashes of light (visual hallucinations). There was no history of prior seizure activity.

An EEG revealed a right temporal lobe spike focus. A CT scan (Fig. 1.1) with intravenous contrast, and an MRI study, were negative. The patient was having visual hallucinations at the time of the CT and MRI scans, and also at the time of the SPECT study. The SPECT study (Fig. 1.2) in the transaxial (A), coronal (B), and sagittal (C) planes revealed a focus of intense, increased activity in the right occipital and posterior temporal lobes (arrows).

Teaching Point:
A focus of intense, increased tracer deposition is typical of an epileptogenic focus in the ictal phase. The cerebral SPECT study may be very useful in defining such foci, particularly when both the CT and MRI scans are negative.

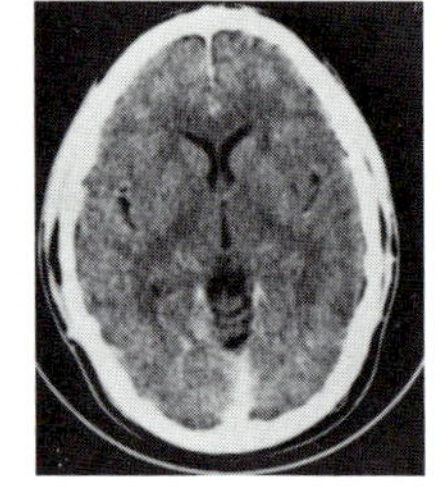
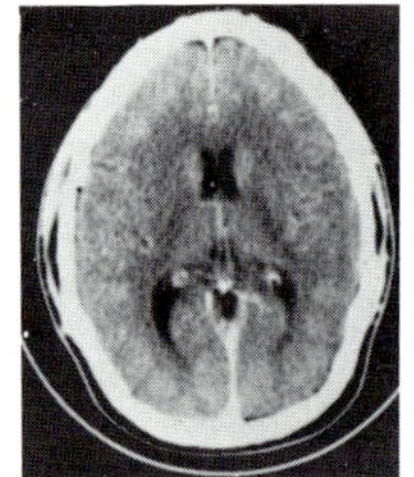

1.1. CT scan.

1.2. SPECT study. A: Transaxial; B: Coronal; and C: Sagittal planes.

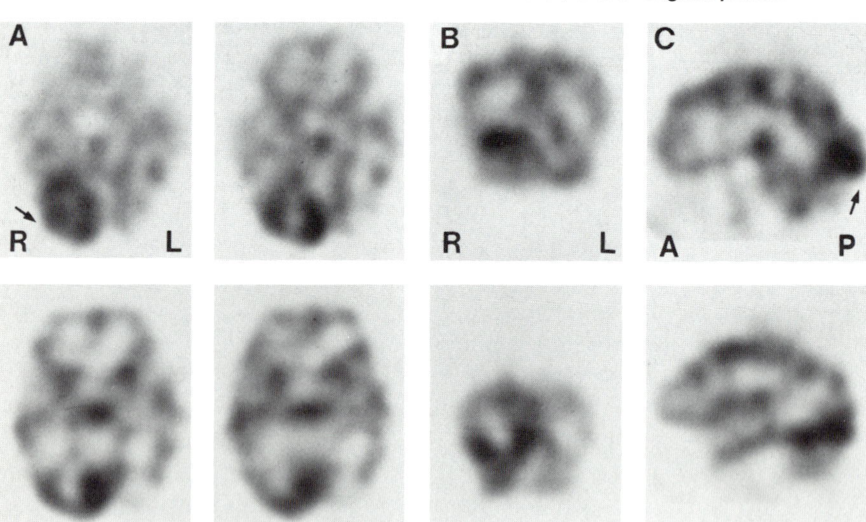

CASE 2: *Contributed by Thomas C. Hill, M.D., SPECT studies acquired on Strichman SME—810.*

This 94-year-old woman was referred for evaluation of status epilepticus. Her seizure activity began eight hours before her hospital admission. There was no history of a seizure disorder.

An emergency CT scan (Fig. 2.1) was unremarkable except for a benign calcification in the left temporal lobe. An EEG showed continuous bursts of left frontal temporal seizure activity. At the time of admission the patient was started on intravenous phenytoin, which was continued over the next two days; she slowly became more responsive but was still disoriented. Four days after admission, a cerebral SPECT study (Fig. 2.2) in the transaxial plane showed an intense focus of increased tracer deposition in the left frontal lobe (arrowhead). As a result, the patient's medication was adjusted and she became more oriented and alert.

Teaching Point:
In this case, the cerebral SPECT study was very helpful in determining the presence of persistent seizure activity despite what was thought to be adequate dosage levels of anticonvulsant medication.

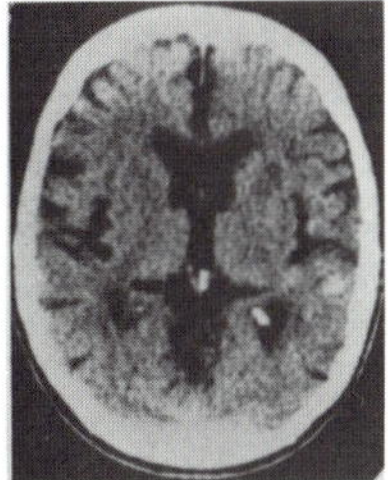

2.1. CT scan.

2.2. SPECT study in transaxial plane.

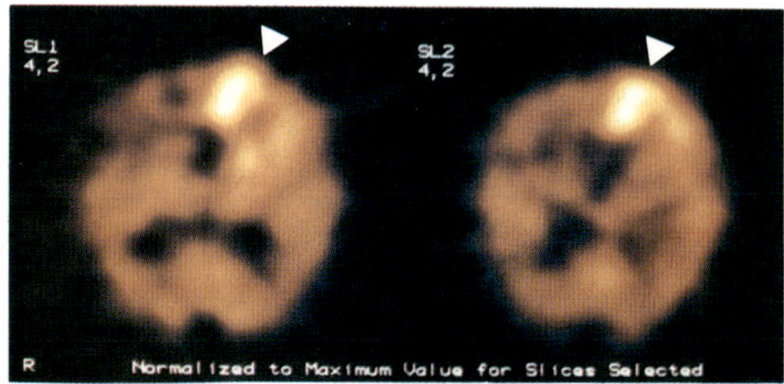

CASE 3: SPECT studies acquired on GE 400AC/T; STAR.

This 41-year-old, right-handed woman was referred for evaluation of a new onset of seizures after a long history of migraine headaches. A CT scan was normal; a MRI study (Fig. 3.1) showed increased signal intensity in the medial aspect of the right temporal lobe, without associated edema or mass effect. An EEG was abnormal with a significant slow wave in the right temporal lobe. A cerebral SPECT study (Fig. 3.2) in the transaxial (A), coronal (B), and sagittal (C) planes was done in the immediate postictal period, while the patient was still complaining of generalized headaches. The SPECT study showed a significant increase in tracer deposition in the right temporal, posterior frontal, and anterior parietal lobes and in the adjacent subcortical areas (arrows).

A follow-up cerebral SPECT study (Fig. 3.3) performed eight days after the first study, showed a normal pattern of tracer distribution.

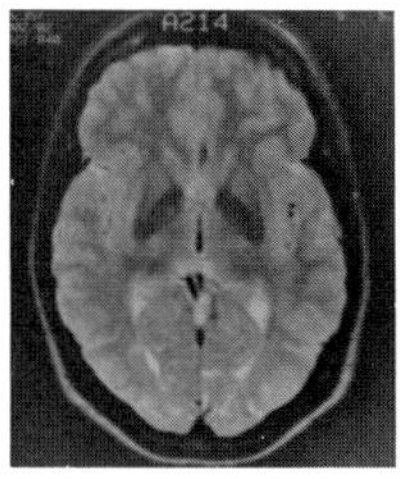

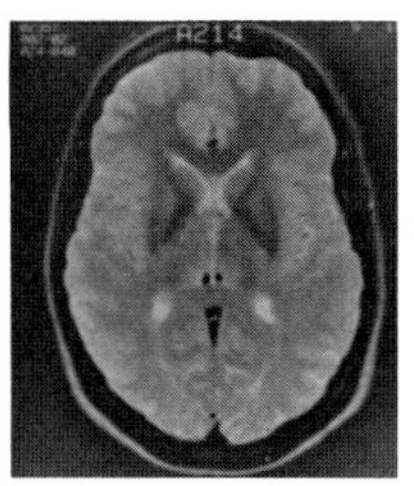

3.1. MRI study.

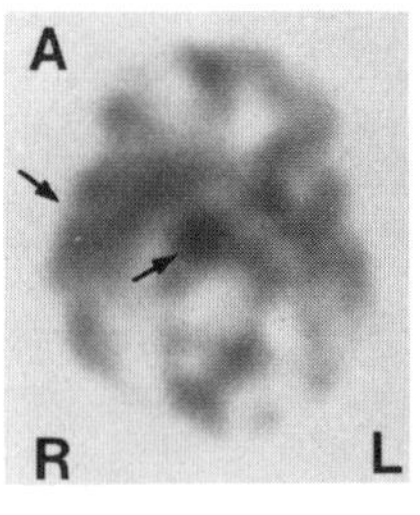

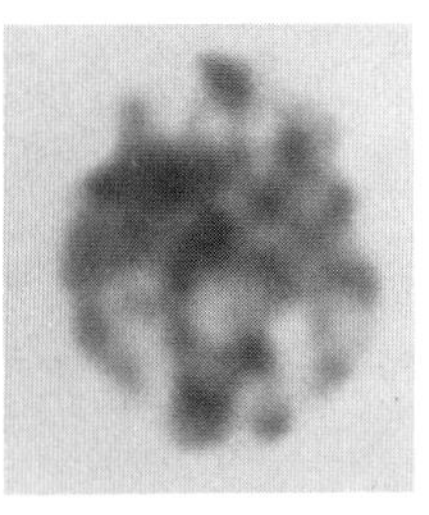

3.2. Initial SPECT study. A: Transaxial; B: Coronal; and C: Sagittal planes.

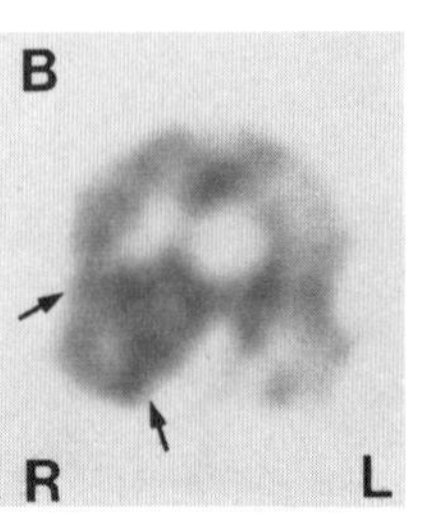

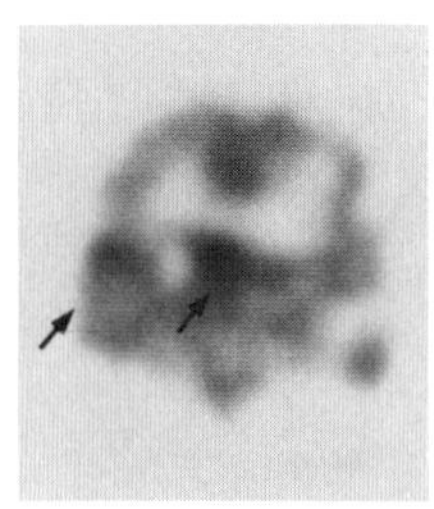

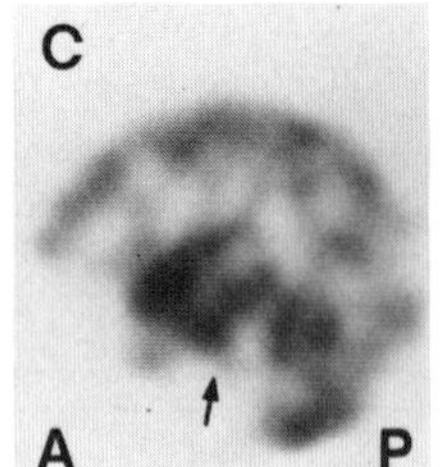

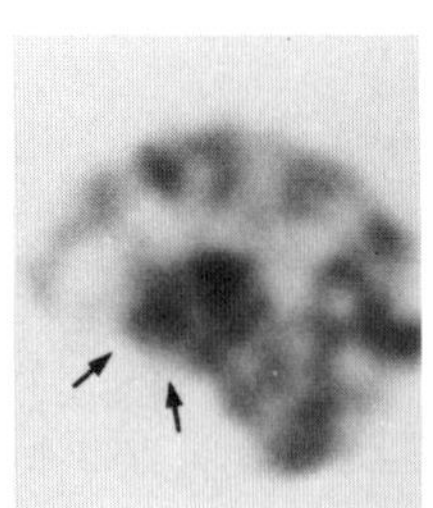

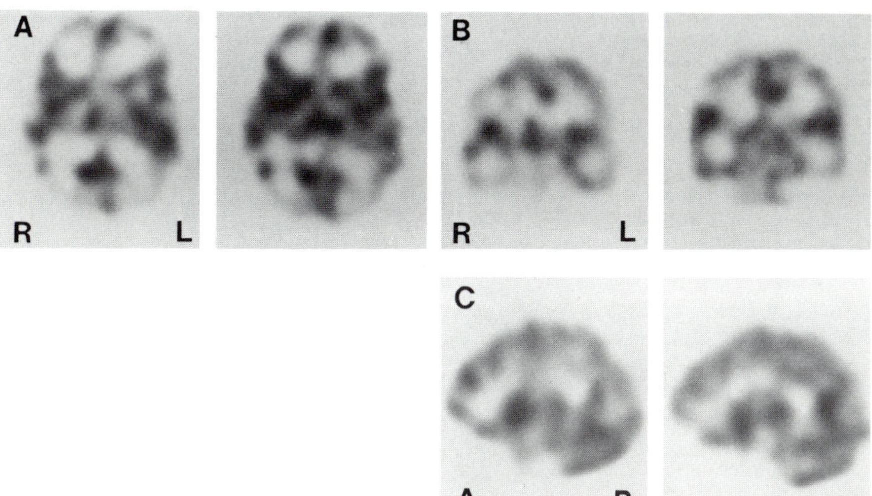

3.3. Follow-up SPECT study.

Teaching Point:
The cerebral SPECT study is frequently useful in localizing and monitoring the seizure activity of epileptogenic foci.

CASE 4: *Contributed by Amjad Ali, M.D., SPECT studies acquired on Siemens 7500 ZLC; VAX Computer.*

This 19-year-old woman had a known seizure disorder since she was nine months old. She was referred for an evaluation before surgical removal of the ictal focus. Prolonged EEG monitoring showed prominent epileptiform activity in the right anterior temporal region. PET scanning with F-18,2,deoxyglucose showed hypometabolic activity in the right temporal lobe. The CT and MRI scans (Fig. 4.1) were normal.

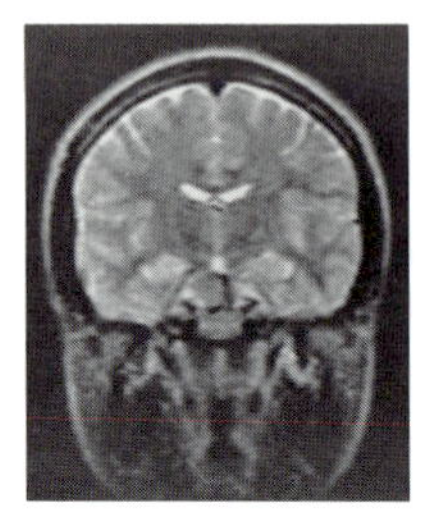

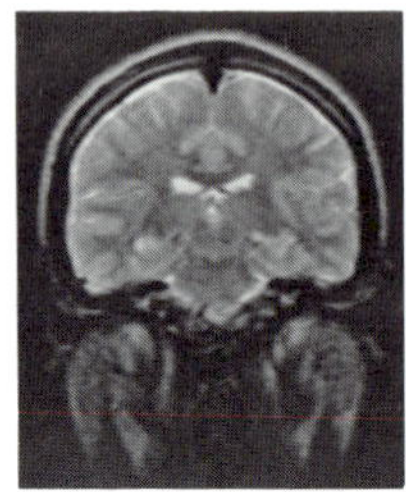

4.1. MRI study.

A cerebral SPECT study (Fig. 4.2) in the transaxial (A) and coronal (B) planes showed a marked decrease in tracer deposition in the right temporal lobe (arrows), corresponding to the abnormality found on the PET scan and the EEG. The patient underwent a resection of the right temporal lobe and was reported to be doing well at a six-month post-surgical follow-up.

4.2. SPECT study. A: Transaxial and B: Coronal planes.

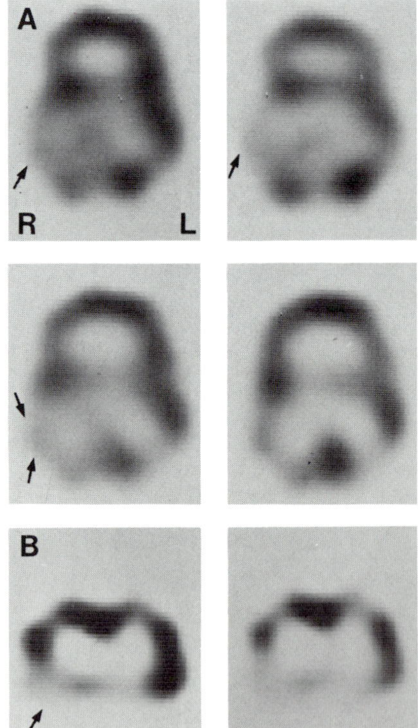

Teaching Point:
As this case illustrates, a cerebral SPECT study done during the interictal phase may be quite useful for the accurate presurgical localization of an epileptogenic focus.

CASE 5: *Contributed by Amjad Ali, M.D., SPECT studies acquired on Siemens 7500 ZLC; VAX Computer.*

A 22-year-old man with a history of epilepsy since the age of nine years was referred for the presurgical evaluation of an epileptogenic focus.

Surface EEG showed frequent bursts of epileptiform activity in both posterior frontal regions, with slightly greater predominance of activity on the right side. A subsequent EEG done during a WADA test showed similar findings. A methohexitol test demonstrated epileptiform foci in the left frontal anguli gyrus and the left temporal lobe.

A PET scan with F-18,2,deoxyglucose demonstrated hypometabolism in the left posterior temporal and inferior parietal lobes. A follow-up MRI showed increased signal activity in the left posterior temporal—parietal region.

A cerebral SPECT study in the transaxial plane (Fig. 5.1) revealed a decreased tracer deposition in the left temporal lobe and adjacent posterior parietal region (arrows).

The patient underwent resection of the left anterior temporal lobe and the inferior tip of the left frontal lobe. Six months after surgery, he was reported to be doing well, without any recurrence of seizures.

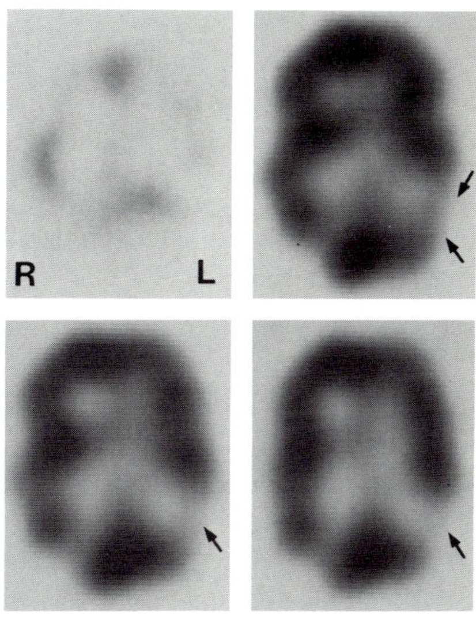

5.1. SPECT study in transaxial plane.

SECTION II: D

Psychiatric Disorders

INTRODUCTION

The major psychiatric disorders—the schizophrenias and the affective disorders—certainly involve alterations of normal brain mechanisms. This has always been accepted, but the pathophysiology of these disorders has not been determined. The term ''functional,'' as opposed to ''organic,'' in describing schizophrenia and manic-depression, was originally used to indicate pathology of function in the absence of gross or microscopic evidence of brain disease. Over the past few years, there have been several methods introduced that show great promise for studying neurobiologic mechanisms related to psychiatric disorders. These new methods include computed tomography (CT), magnetic resonance imaging (MRI), techniques of molecular genetics, tomographic brain mapping, functional brain imaging using positron emission tomography (PET), and single photon emission computed tomography (SPECT). SPECT imaging shows great potential for helping in the understanding of the pathophysiology of these disorders.

Psychiatric diagnosis remains essentially a product of the history and physical examination of the patient. There are as yet no specific biological markers for psychiatric diagnoses. The mental status examination is the psychiatric equivalent of the physical examination and requires the collecting of signs and symptoms in a systematic manner. The recent introduction of the third edition of the ''Diagnostic and Statistical Manual of Mental Disorders'' (DSM III) has significantly increased the reliability of psychiatric diagnosis by introducing operationalized diagnostic criteria. However, basic questions remain about the validity of many psychiatric diagnostic categories. Most clinicians and researchers today agree that the major diagnostic categories are probably heterogeneous: they present with similar clinical pictures, but are composed of different subtypes. A nonpsychiatric medical analogy would be the current recognition of at least two subtypes of diabetes mellitus. New methods for looking at brain mechanisms, such as SPECT, may reveal differences within diagnostic categories and possibly give leads to treatment response and prognosis.

In addition to the problems of heterogeneity within diagnostic categories, there are problems of differential diagnosis in psychiatry. In general, psychopathologic signs and symptoms are not specific to a particular diagnosis. For instance, hallucinations are found in schizophrenia, mania, and organic brain syndromes; cognitive impairment is common in dementia and severe depression (the so-called ''pseudodementia''). Techniques such as PET and SPECT may be of help in the differential diagnosis.

Schizophrenias (Cases 1-4)

CT studies have shown ventricular enlargement in schizophrenic patients when compared to matched controls. These findings are not specific for schizophrenia: they have also been noted in manic and depressed patients, and they are not found in all schizophrenics.

MRI studies which provide greater resolution and detail than CT, also show ventricular enlargement in some schizophrenics and a decreased frontal lobe size (''hypofrontality''). In several studies, PET also has demonstrated hypofrontality in schizophrenics, but these results have not been consistently reproduced. PET studies of basal ganglia metabolism in schizophrenics have been contradictory, with some studies showing a decrease and some an increase. PET is a complex and expensive method, so that, while the resolution with SPECT is not quite as good as with PET, SPECT is a more feasible procedure in most clinical settings. SPECT studies using I^{123} IMP with schizophrenics are in the early stages of development. However, initial results show hypofrontality in some but not all patients. The most consistent finding has been an increased rCBF in the basal ganglia—specifically, the caudate, which is associated with positive symptoms such as delusions and hallucinations. Some acute schizophrenics also have increased rCBF in the temporal lobes. The specificity of these findings and the implications for predicting response to treatment need further study.

Affective Disorders (Cases 5-6)

Brain imaging studies in the affective disorders have been less extensive to date. Ventricular enlargement has been noted in bipolar disorders and depressions: PET studies have shown overall increased cortical metabolism when patients are in the manic phase and decreased cortical metabolism in the same patients in the depressed phase. Initial SPECT studies using I^{123} IMP in patients with major depressive disorders show an overall decrease in rCBF especially in the frontal lobes. This pattern is different from that seen in the dementias, and it may provide a method for the differential diagnosis of dementia and the pseudodementia of depression. The decreased rCBF seen in depressed patients appears to normalize as the patient improves clinically.

CASE 1: *SPECT studies acquired on GE 400AC/T; STAR II.*

This 34-year-old homeless woman was referred for evaluation, while having auditory hallucinations and paranoid delusions. The clinical diagnosis was paranoid schizophrenia, which was being treated with fluphenazine, 10 mg bid.

The cerebral SPECT study (Fig. 1.1) in the transaxial (A), coronal (B), and sagittal (C) planes showed increased tracer deposition in the caudate nuclei (arrowheads) and the right posterior temporal region (arrows).

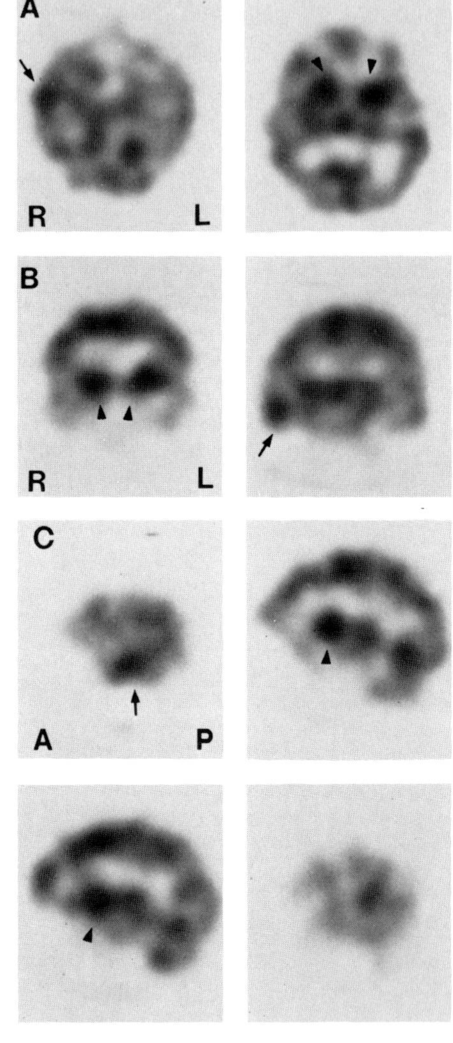

1.1. SPECT study. A: Transaxial; B: Coronal; and C: Sagittal planes.

CASE 2: *SPECT studies acquired on GE 400AC/T; STAR II.*

This 35-year-old man, who was known to be paranoid schizophrenic, was referred for evaluation of his progressive paranoid delusions and auditory hallucinations. The patient had stopped taking his medication one month before admission to the hospital. A CT scan (Fig. 2.1) was negative.

Before the cerebral SPECT study, the patient had been started on fluphenazine, 10 mg bid, and benztropine, 1 mg bid. The cerebral SPECT study (Fig. 2.2) in the transaxial (A), coronal (B), and sagittal (C) planes revealed increased tracer deposition in the caudate nuclei (arrows) and slightly increased tracer activity in the right supratemporal region (arrowhead).

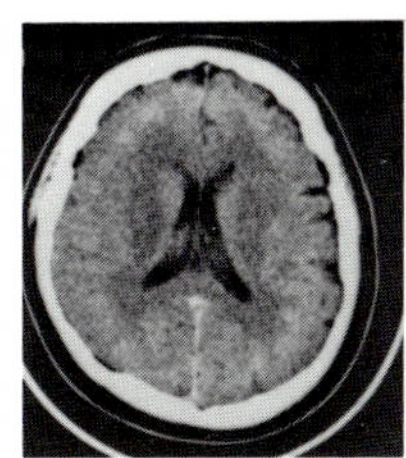

2.1. CT scan.

2.2. SPECT study. A: Transaxial; B: Coronal; and C: Sagittal planes.

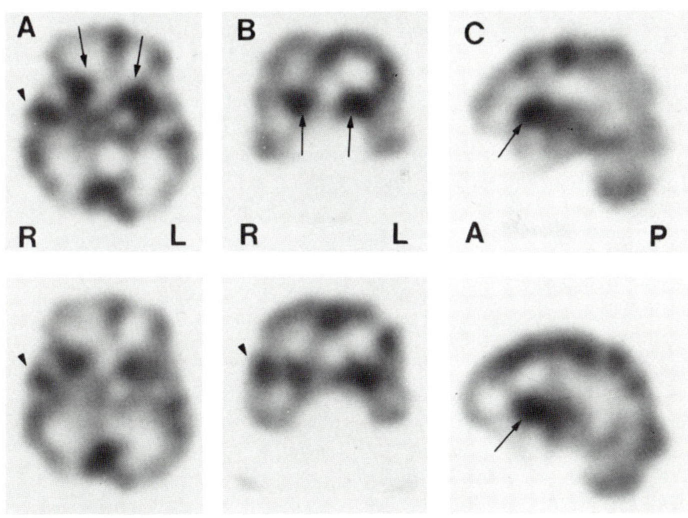

CASE 3: *SPECT studies acquired on GE 400AC/T; STAR II.*

This 30-year-old man was referred for evaluation of an acute psychosis characterized by paranoid delusions, auditory hallucinations, and destructive behavior that had become progressively worse over the preceding two years. An EEG and CT scan were normal. The clinical diagnosis was acute paranoid schizophrenia.

At the time of the initial SPECT study, the patient was receiving a total daily dose of 100 mg of loxapine (an antipsychotic medication). The initial cerebral SPECT study (Fig. 3.1) in the transaxial (A), coronal (B), and sagittal (C) planes revealed an increased tracer deposition in the caudate nuclei (arrows), along with a mild decrease of tracer activity in the cortex of the frontal lobes (arrowheads). Two weeks later, a follow-up cerebral SPECT study was performed. At that time, the patient was much improved clinically. The follow-up study (Fig. 3.2) demonstrated a significant but partial decrease of tracer activity in the caudate nuclei. A slight focal increase in tracer deposition was noted in the right temporal lobe (arrows). This later finding may have been related to the patient's medication lowering the seizure threshold.

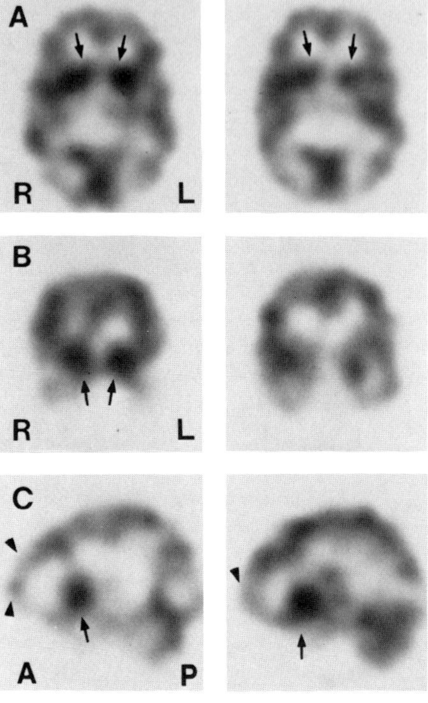

3.1. Initial SPECT study. A: Transaxial; B: Coronal; and C: Sagittal planes.

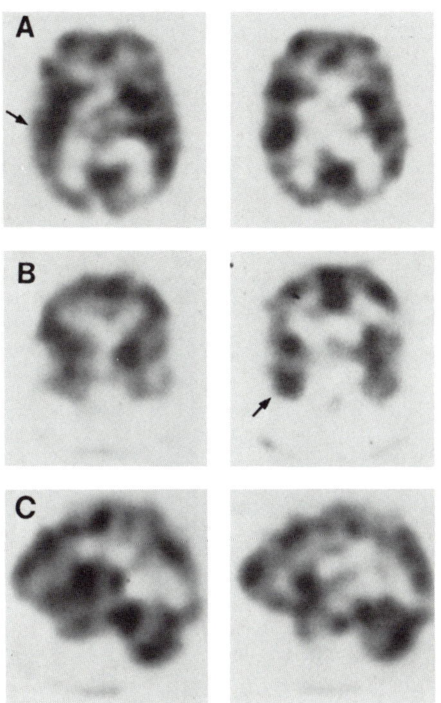

3.2. Follow-up SPECT study.

Teaching Point:
As the preceding three cases illustrate, the pattern of increased tracer activity in the caudate nuclei (with or without decreased tracer uptake in the frontal lobe) is characteristically seen in patients with acute psychoses.

CASE 4: SPECT studies acquired on GE 400AC/T; STAR II.

This 41-year-old woman, known to be paranoid schizophrenic, was referred for evaluation of threatening auditory hallucinations. An EEG and a CT scan (Fig. 4.1) were normal. At the time of the cerebral SPECT examination, she had been receiving trifluoperazine, 50 mg daily, and benztropine, 1 mg daily.

The cerebral SPECT study (Fig. 4.2) in the tansaxial (A) and sagittal (B) planes revealed increased tracer activity in the caudate nuclei (arrows), with focal increased uptake of tracer in the right temporal lobe (arrowhead). As a result of her continuing hallucinations and the cerebral SPECT finding of increased temporal lobe activity, the patient's medication was modified to include an anticonvulsant (carbamazepine). Following modification of her medication, the patient showed a dramatic improvement, with a marked decrease in the frequency of auditory hallucinations.

Teaching Point:
As this case demonstrates, the cerebral SPECT study may reveal useful information that is not demonstrable on other studies such as the EEG or CT scan.

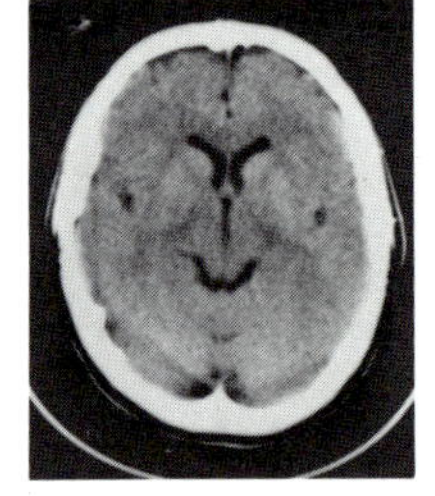
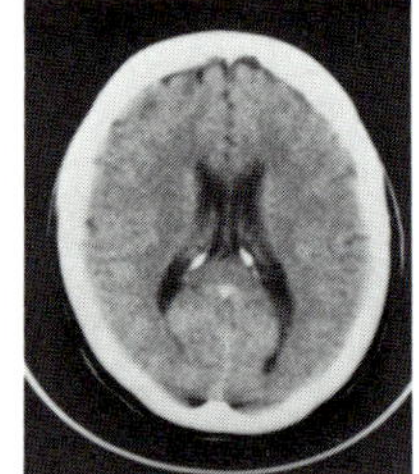

4.1. CT scan.

4.2. SPECT study. A: Transaxial and B: Sagittal planes.

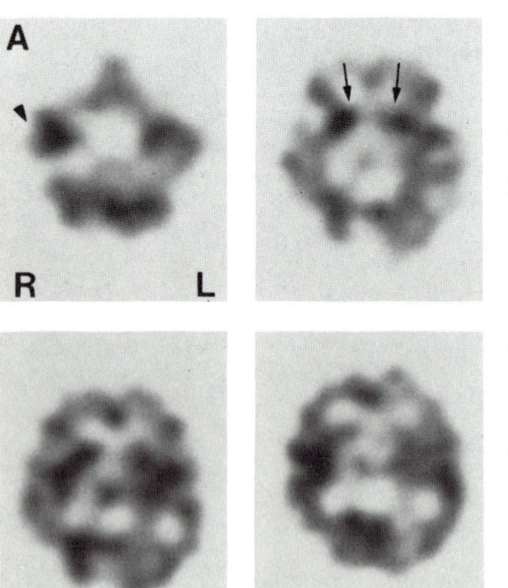

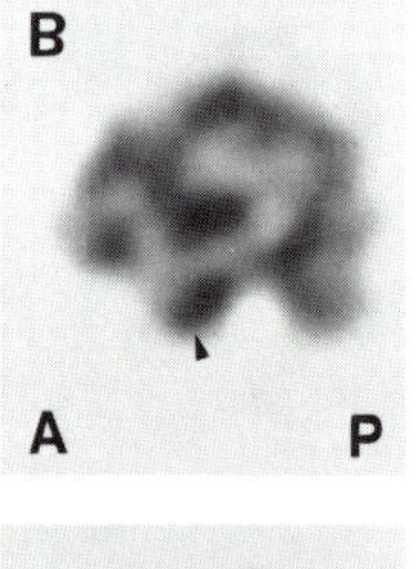

CASE 5: *SPECT studies acquired on GE 400AC/T; STAR II.*

Following an attempted suicide, this 32-year-old man was referred for evaluation of severe depression. An EEG and CT scan (Fig. 5.1) were normal. At the time of the cerebral SPECT study, the patient was not on medication.

The cerebral SPECT study (Fig. 5.2) in the transaxial (A), coronal (B), and sagittal (C) planes revealed an overall decrease in tracer deposition throughout the cerebral cortex. This pattern was most marked in the frontal lobes (arrows).

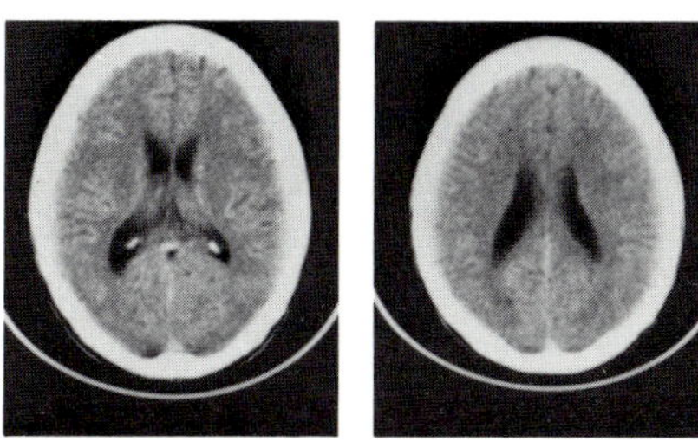

5.1. CT scan.

5.2. SPECT study. A: Transaxial; B: Coronal; and C: Sagittal planes.

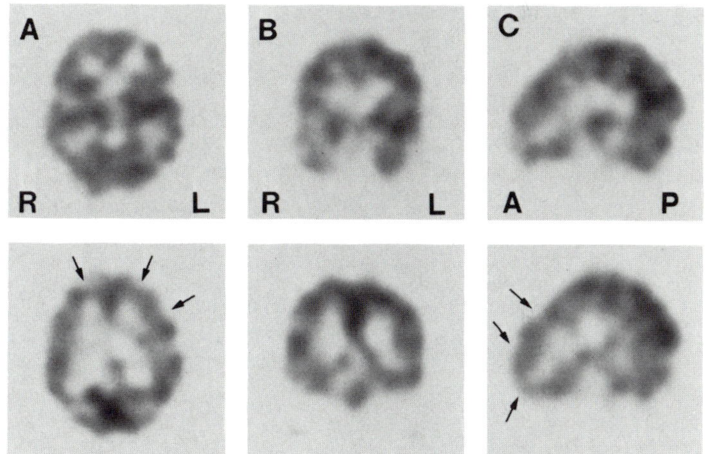

CASE 6: *SPECT studies acquired on GE 400AC/T; STAR II.*

This 56-year-old man, with a history of major depressive disorder, was referred for evaluation of a severe depression with paranoid ideation but no hallucinations. The CT scan (Fig. 6.1) and EEG were normal. At the time of the cerebral SPECT study, the patient was receiving amitriptyline, 50 mg daily.

The cerebral SPECT study (Fig. 6.2) in the transaxial (A), coronal (B), and sagittal (C) planes revealed a decrease in tracer deposition most marked in the frontal lobes (arrows).

Teaching Point:
As the preceding two cases illustrate, the pattern seen in major depressive disorders is quite different from that noted in acute psychoses such as schizophrenia. The relative decrease of tracer uptake appears to correlate with the severity of the depression.

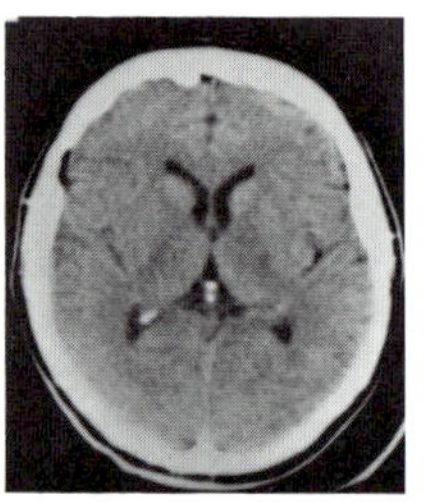

6.1. CT scan.

6.2. SPECT study. A: Transaxial; B: Coronal; and C: Sagittal planes.

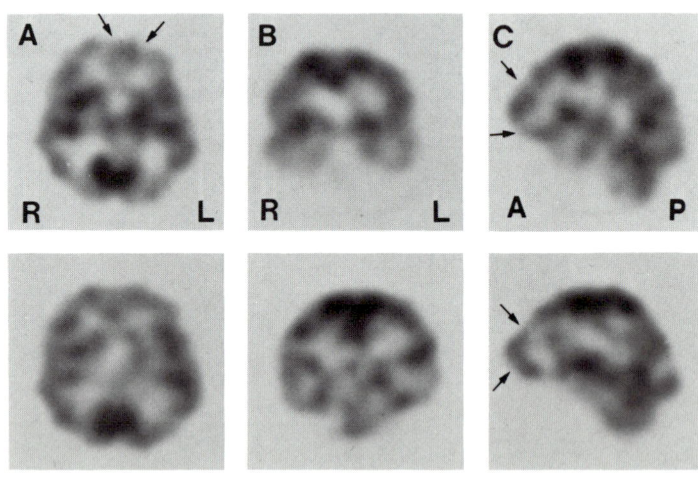

SECTION II: E

Miscellaneous Disorders

INTRODUCTION

In addition to cerebrovascular disease, dementia, epilepsy, and psychiatric illness, there are a variety of other disease states in which cerebral SPECT imaging may prove useful. These disorders include head trauma, anoxic states, and tumors.

CASE 1: SPECT studies acquired on GE 400AC/T; STAR.

This 23-year-old man was referred for evaluation of a left hemiparesis and loss of cognitive function after sustaining a gunshot wound to the head. An initial CT scan showed hyperdense hematomas in the right frontal and deep right parietal lobes, which appeared hypodense on a follow-up CT scan.

A cerebral SPECT study (Fig. 1.1) in the transaxial (A), coronal (B), and sagittal (C) planes revealed a marked decrease in tracer deposition throughout the right cerebral hemisphere, plus an absence of tracer activity in the right basal ganglia and the frontal lobes; tracer activity was less reduced in the right parietal lobe. The defect shown by SPECT study was of a greater area than that noted on the CT scan. The pattern of abnormal tracer deposition appreciated by the SPECT study clearly delineated the trajectory of the bullet (arrows).

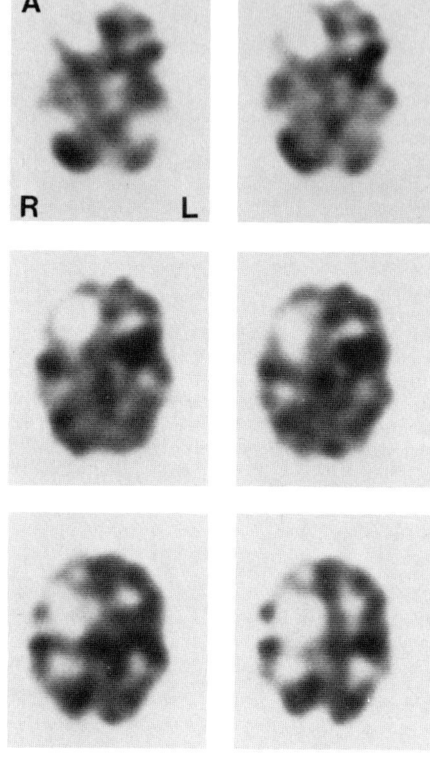

1.1. SPECT study. A: Transaxial; B: Coronal; and C: Sagittal planes.

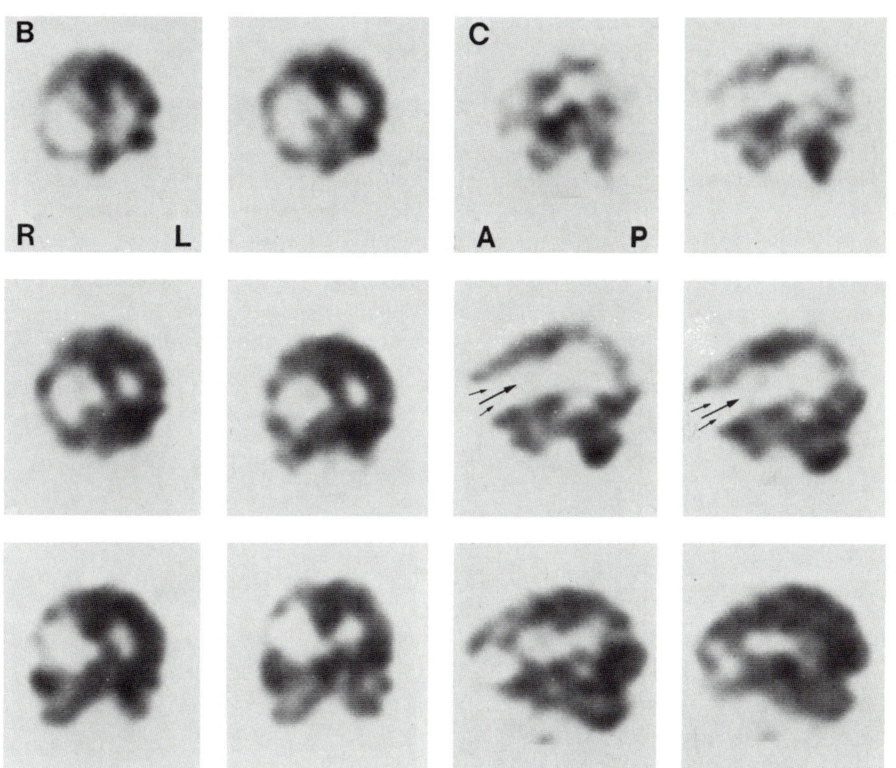
B
R
L
C
A
P

CASE 2: SPECT studies acquired on GE 400AC/T; STAR.

This 23-year-old man was referred for the evaluation of closed intracranial trauma sustained during an auto—train collision. At the time of evaluation, the patient was in a Grade II coma. An initial CT scan (Fig. 2.1) revealed a small left temporoparietal epidural hematoma, a small midline hemorrhage in the third ventricle, and a large extracranial scalp hematoma overlying the left frontal temporoparietal calvarium. A follow-up CT scan (Fig. 2.2), one month later, showed resolution of the areas of hemorrhage, a low-density area in the upper mesencephalon, and some enlargement of the ventricles and sulci (the result of resolution of the cerebral edema).

A cerebral SPECT study (Fig. 2.3) in the transaxial plane showed an overall heterogeneous pattern of tracer distribution, with a lack of normal delineation between the cortical gray and white matter. This was particularly pronounced in the frontal, temporal, and parietal cortex. There was also very poor delineation of subcortical structures.

These findings suggest a gross disorganization of regional cerebral blood flow and metabolism.

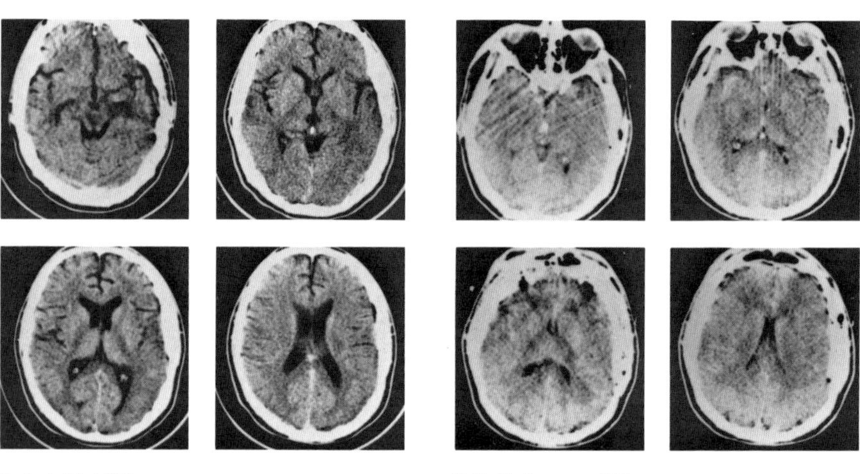

2.1. Initial CT scan.

2.2. Follow-up CT scan.

2.3. SPECT study in transaxial plane.

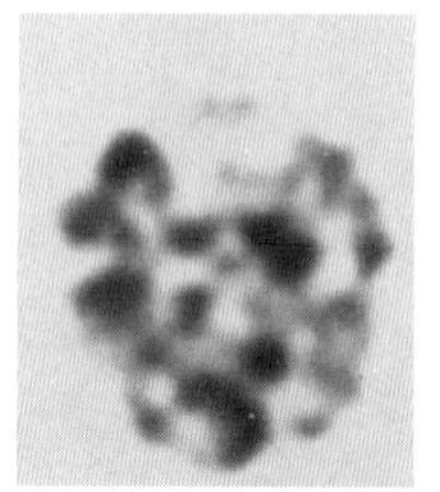
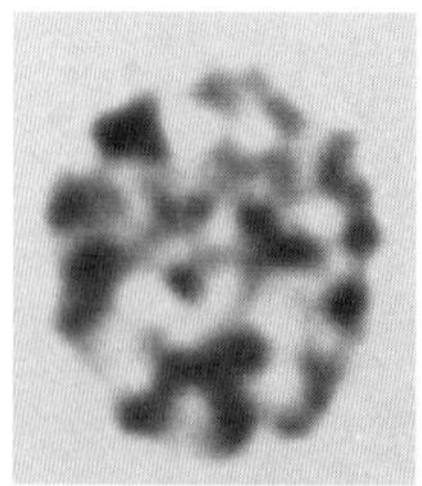
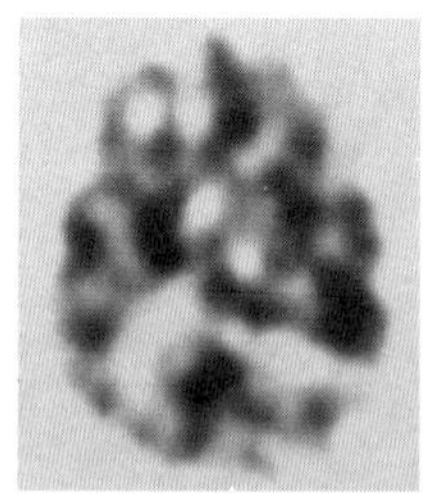
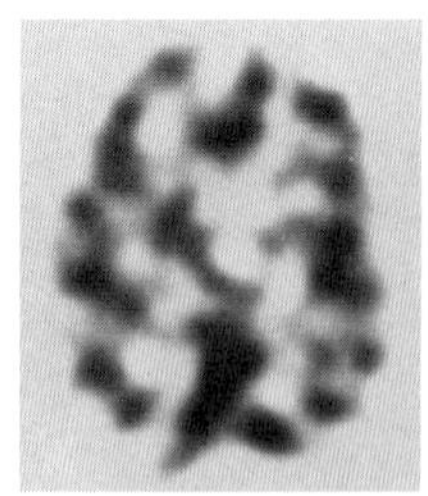

***CASE 3:** Contributed by Joji Nagawara, M.D., SPECT studies acquired on Shimadzu HEADTOME SET-031.*

This 65-year-old man was referred for evaluation of right-sided headaches and a left hemiparesis. Two months previously, he had sustained head trauma. On a CT scan, a subdural hematoma was noted on the right side.

An initial cerebral SPECT study (Fig. 3.1 A) in the transaxial plane revealed decreased peripheral tracer deposition in the right cerebral hemisphere (arrows). In addition, crossed cerebellar diaschisis was observed (arrowhead).

A week after the subdural hematoma was drained, a follow-up cerebral SPECT study (Fig. 3.1 B) showed a significantly improved tracer deposition in the periphery of the right cerebral hemisphere (arrows).

3.1A. Initial SPECT study in transaxial plane.

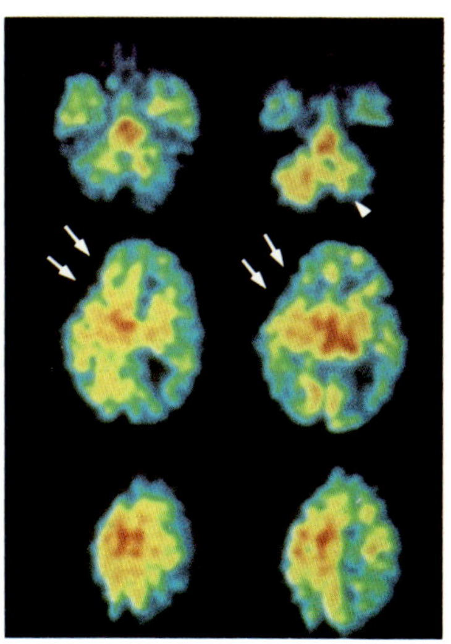

3.1B. Follow-up SPECT study.

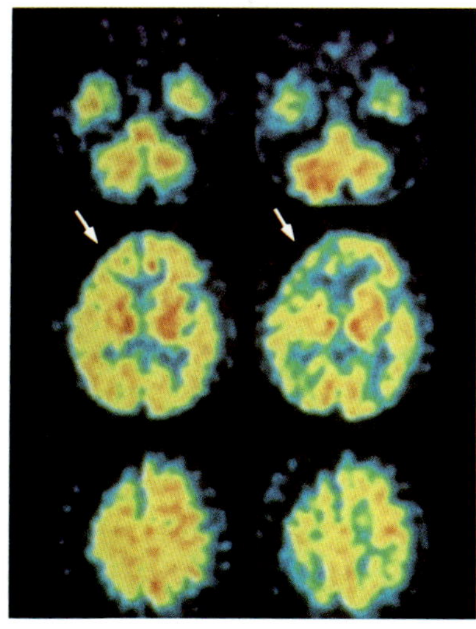

CASE 4: *SPECT studies acquired on GE 400AC/T; STAR II.*

This 76-year-old woman at the time of a cardiac arrest, sustained an anoxic encephalopathy. Despite vigorous attempts at cardiopulmonary resuscitation, the patient remained severely obtunded at the time of referral.

The cerebral SPECT study (Fig. 4.1) in the transaxial (A), coronal (B), and sagittal (C) planes revealed a marked decrease in tracer deposition throughout the cerebral cortex. This was most marked in the occipital, parietal, and frontal lobes as well as the periventricular white matter. The brain stem and cerebellum were relatively spared.

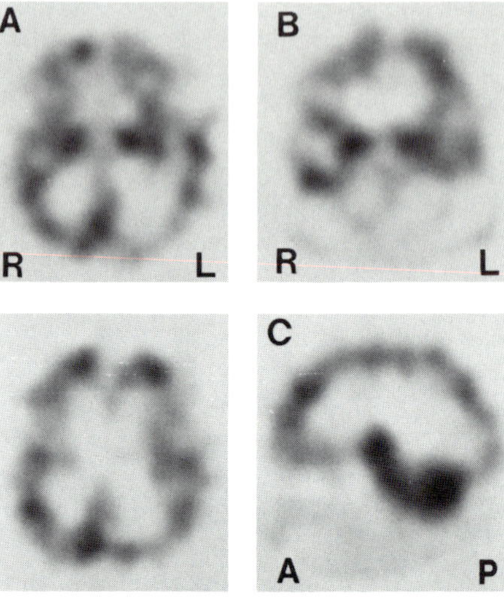

4.1. SPECT study. A: Transaxial; B: Coronal; and C: Sagittal planes.

Teaching Point:
This pattern is typically seen in patients who have sustained a severe anoxic episode.

CASE 5: SPECT studies acquired on GE 400AC/T; STAR.

This 42-year-old woman was referred for evaluation of vasospasm sustained at the time of a vertebral (bilateral) artery angiogram. A MRI study (Fig. 5.1) showed an inhomogeneous mass in the right cerebellopontine cistern with compression and displacement of the brain stem, the right cerebellar hemisphere, and the fourth ventricle.

A cerebral SPECT study (Fig. 5.2) in the transaxial plane showed tracer deposition was absent at the site of the right cerebellopontine mass (arrowhead). A second site of decreased tracer deposition was noted (arrow) in the right occipital lobe. It was thought that this occipital region finding represented an ischemic zone.

Teaching Point:
In this case, the SPECT study was quite helpful in explaining the patient's symptoms, which clinically did not appear to be related to the acoustic neuroma noted on the MRI study.

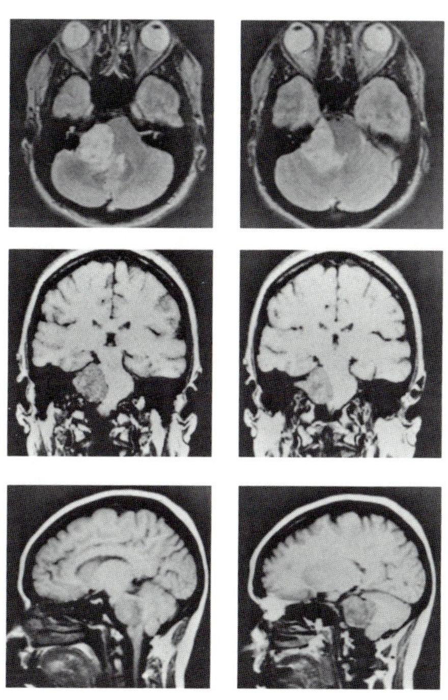

5.1. MRI study.

5.2. SPECT study in transaxial plane.

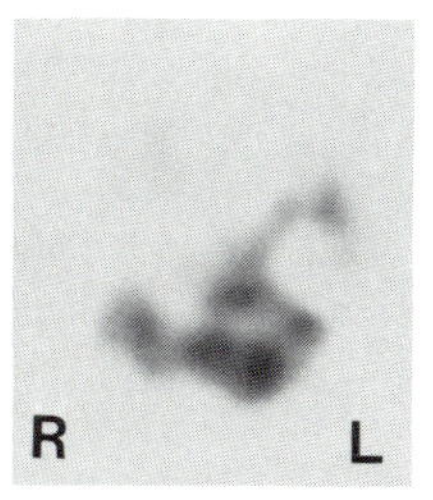

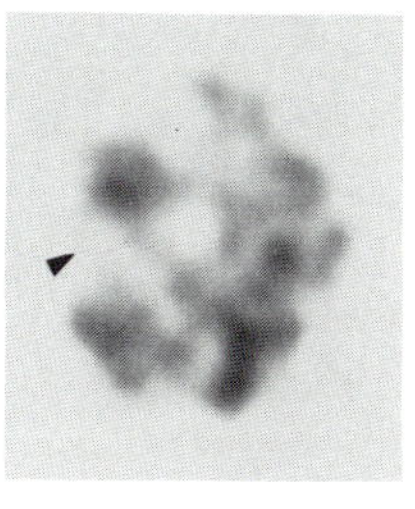

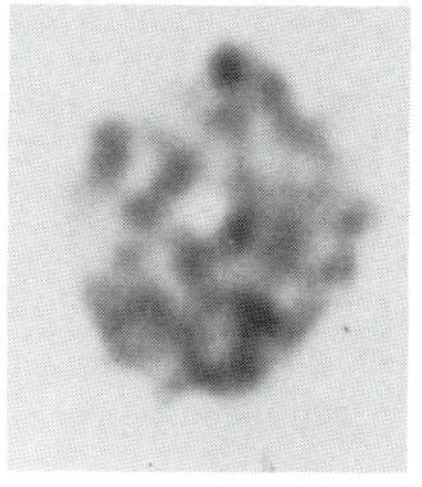

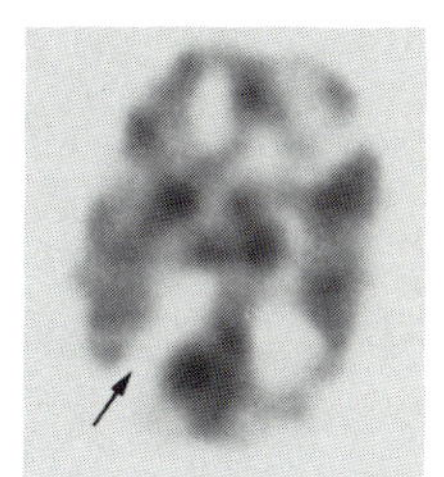

CASE 6: *Contributed by Donald Fox, M.D. and Caesar Mayo, M.D., SPECT studies acquired on Spectrum 150DT: SOPHA Computer.*

This 61-year-old man was referred for evaluation of increasing confusion. Previously, he had undergone a craniotomy for a cystic astrocytoma of the left temporal lobe. The surgery was followed by radiation and chemotherapy three years prior to his referral for a SPECT study.

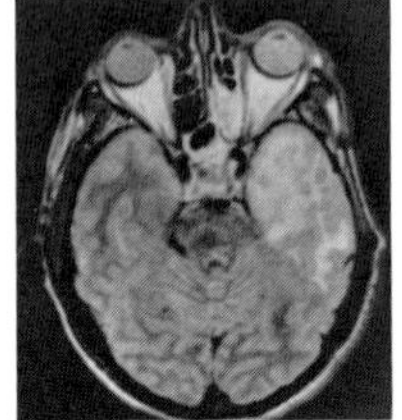

6.1. MRI study.

The last follow-up MRI study (Fig. 6.1) showed a diffuse increase in signal intensity throughout the left temporal lobe. This finding was thought to be most consistent with progressive changes attributable to the radiation therapy.

A cerebral SPECT study (Fig. 6.2) in the transaxial (A), coronal (B), and sagittal (C) planes showed decreased tracer deposition in the left temporal lobe (arrows). The findings were thought to be consistent with ischemic changes in the temporal lobe, possibly secondary to radiation therapy.

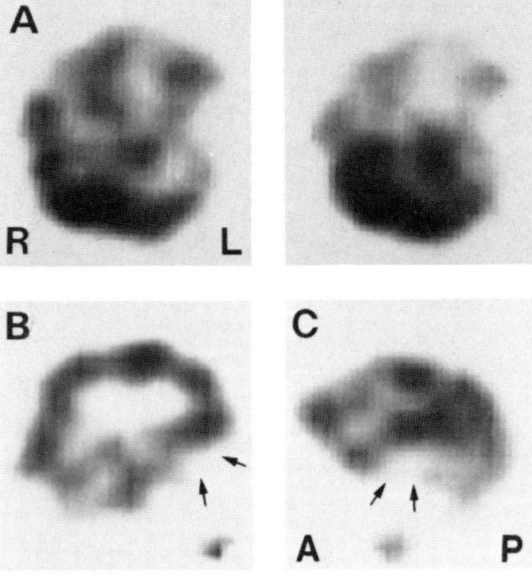

6.2. SPECT study. A: Transaxial; B: Coronal; and C: Sagittal planes.

SECTION III:

Appendix

Quality Control

SPECT Quality Control Procedures

RADIOPHARMACEUTICAL PROCEDURES

Radiochemical Purity Checks:

a. Not required, but helpful to perform on each batch of radiopharmaceuticals received
b. Procedure uses ascending thin-layer chromatography (ITC) with a silica gel chromatography strip and a methanol, chloroform, glacial acid (18/15/1 V/V/V) solution

Note: This procedure requires only a short time. The silica gel strips are cut in half and counted in a dose calibrator or well-counter chamber. The percentage of free I^{123} is calculated as follows:

$$\% \text{ free } I^{123} = \frac{\text{counts strip \#1 (top)}}{\text{counts strip \#1 \& strip \# 2 x 100}}$$

The free I^{123} component should not exceed 5%

Radionuclide Purity:

a. Need not be checked on a regular basis when using the (p,5n) reaction product
b. This product should not contain less than 98% I^{123} at the time of calibration

Radiopharmaceutical Dose: Each dose should be checked on a dose calibrator to confirm the calculated radioactivity before IV injection

Radiopharmaceutical Solution: The solution should be visually inspected for particulate matter and discoloration before administration

INSTRUMENTATION PROCEDURES

Weekly QC for SPECT Cameras:
a. 30 million count extrinsic flood
b. 3 million count intrinsic flood

Weekly Uniformity Flood:
a. 30 to 100 million counts
b. Co^{57} or Tc^{99m} flood sources
c. Extrinsic

Assuring Flat Uniformity:
a. Take two high-count floods, rotate second 180 degrees from the first
b. Correct the second flood with the first
c. Methods
   1. visual inspection
   2. profiles
   3. quantification of relative standard deviation

Note: Uniformity correction should be performed with the gantry positioned at 0°, 90°, 180°, and 270°.

Center of Rotation:
a. Offsets must be less than ± 0.6 pixels
b. Recalculate daily or whenever camera field changes

Daily QC for SPECT Cameras:
a. 3 million count for extrinsic flood
b. center of rotation
c. QC during acquisition
d. QC after acquisition

GENERAL PROCEDURES FOR QUALITY CONTROL

During acquisition:
a. Try to keep the patient from moving
b. The camera *must be absolutely parallel* to the axis of rotation

After acquisition:
a. Pseudo—three-dimensional display (cine playback of views) to detect patient motion and camera error
b. Sinogram—based on projection data in a single transverse slice, all laid on top of one another to form a picture.

Equipment Protocols

ADAC Laboratories

Patient Preparation:
a. 200mg potassium perchlorate
b. Eyes closed, image 25 minutes

Acquisition:
a. ARC 3000 Gamma Camera
   1. slant hole collimator
   2. multiple energy acquisition/scatter correction
   3. 360 ° rotation
   4. 64 x 64 x 16 matrix
   5. 64 azimuths
   6. 30 seconds per azimuth
   7. 16-cm radius

Processing:
a. Gaussian filter
   1. cutoff: 0.4
   2. filter order: 25
   3. scale factor: 5
b. Low pass filter: 0.4 Nyquist
c. High speed reconstruction: Hamming filter cutoff of 10
d. Apply changes attenuation correction

ELSCINT

Patient Preparation:
a. Have low ambient noise and light (eyes closed)
b. Be sure SPECT imaging table is comfortable
c. Administer 3 to 4 mCi dose IV
d. Begin acquisition approximately 15 minutes after injection

Acquisition:
a. Collimator: APC-5 (medium energy) or APC-4
b. 360 ° rotation starting at the angle of −270 ° (to pass posteriorly first)
c. Acquire 60 views (continuously or step-and-shoot) i.e., ang. step = 6 °
d. Time per view: 40 seconds (40 minutes for the complete scan).*

e. Energy: 159 keV $\pm$ 10%
f. Matrix 64 x 64
g. Orbit: circular
h. Zoom 1.5 *and* X, Y offset; 0, −85 (version F1) or zoom 1.0 and *no* X, Y offset (version FO)

Processing:
a. Normalization (half-life −1, special sensitivity map [only in the case of X, Y offset])
b. Prefilter: none
c. Backprojection filter: Hanning (0,1,4)
d. Postfilter: interslice filter ("time-smooth" twice)**
e. Reconstruction zoom: 1.33
f. Attenuation correction coefficient: 0:11

*If acquiring in continuous mode, use slow speed (zero potentiometer setting). This will make the time for the complete scan 30 to 35 minutes.

**Presets are available for this and other processing functions

GENERAL ELECTRIC Medical Systems

Acquisition (General):
a. Complete routine quality control for tomography.
b. Use LEAP collimator.
c. Inject I.V. 3.5 mCi; Iodoamphetamine HC1 123, image starting 15 to 20 minutes later.

Uniformity Correction (Starcam™ 400 AC/T): Peak for Co57 with a Co57 sheet flood source. (More consistency across the field than fillable source; should be purchased and specified as being uniform for SPECT use).

Patient Positioning:

a. Position patient's head so that the canthomeatal line is at a 5 - 10° angle:
   1. This gives comparable CT transaxial orientation.
   2. Proper head positioning when possible is superior to using oblique angle reconstruction.

b. Rotate camera around the patient so as to come as close to the patient as possible:
   1. Talk patient through this part.
   2. For quality control, measure the distance from the camera in the left and right lateral positions to the tip of the patient's nose to approximate ROR (13-15 cm is acceptable and achievable). Check for patient clearance.
   3. When a distance of less than 15 cm cannot be achieved, head angulation should be changed. Achieving the smallest radius of rotation is of the utmost importance.

Computer Setup (Starcam or Star II™):

a. Run ENERGIES protocol to peak for I^{123}
b. Define tomographic acquisition (ATOM command)
c. 64 x 64 matrix
d. Zoom coordinates will be 512,320, 1.6 (This will represent the pan for the 1.6 zoomed image. These pan numbers are carried into reconstruction and are compensated for. Although the COR and uniformity correction are collected at a different pan and zoom, the computer will automatically correct for this).
e. Tomo type: circular stepwise (If doing elliptical, reference elliptical setup procedure).
f. Tomo direction/start angle: start at right lateral position rotating in counterclockwise direction.

PICKER

Patient Preparation:

a. 40μCi/kg injected IV
b. Image 20 minutes after dose (initial scan)
c. Image 4 hours after dose (delayed scan)

Note: Both scans are acquired in a room with very low light

Acquisition Setup:

a. Medium energy, parallel hole collimator
b. 64 x 64 word matrix
c. 64 angles
d. 40 seconds per angle
e. 360° rotation

Reconstruction Protocol:

a. Center of rotation/flood uniformity corrected
b. Prefiltered using the Metz filter
c. Attenuation correction (prereconstruction: Sorenson method)
d. Reconstruction filter = 1 (Equation: $P = [-1/(2\pi A/\iota - 0.25])$)
e. Filtered backprojection
f. Transverse, sagittal, coronal, and oblique slices

SHIMADZU Medical Systems

Preparation:

a. 3 mCi I-123-IMP
b. Collimator: 10 min. for 3 slice
c. Early Image: 20 minutes after IV injection
d. Delayed Image: 5 hours after injection. Scan time is 15 min. for delayed scan.
e. Additional Scan: The second and third scan may be executed for 6 and 9 transaxial images, respectively.

SIEMENS Medical Systems

Patient Preparation:

a. 200 mg potassium perchlorate oral, 30 to 45 minutes before injection
b. Patient's eyes closed for 20 minutes before injection
c. Dose (adult): 3.0 mCi
d. Inject patient (eyes closed)
   1. then wait 20 minutes before scanning
   2. for 4 hour delay image, patient's eyes closed for 20 minutes before scanning

Equipment:

a. Orbiter 3700 DOT or Scintiview/MicroDELTA™
b. Orbiter 7500 DOT or Scintiview/MicroDELTA™
c. 30° slant hole medium energy collimator (300keV)
d. Medium energy parallel hole collimator (300keV)
e. Use of the weighted acquisition module may dramatically increase image contrast

Quality Control:

a. Flood field correction
b. Center of rotation

Acquisition:

a. Multiple energy acquisition/scatter correction
b. 159 keV with 20% window
c. 360° rotation
d. 64 x 64 x 16 matrix
e. 64 views
f. 15-20 seconds per view
g. Total imaging time: 16-21 minutes
h. Use head holder for close acquisition

Processing:

a. Rush Presbyterian—St. Lukes, Chicago, IL (Mark Groch, MS)
   1. Butterworth filter
      - cutoff: 0.35
      - filter order: 10
   2. slice averaging filter
      – 5 adjacent slices
b. VA Minneapolis, Minneapolis, MN (Frank Grund, MD)

1. Butterworth filter
   - cutoff: 0.4
   - filter order: 7
2. slice averaging filter
   – 3 adjacent slices

c. Apply attenuation correction;
   (1) Sorenson—on the fly; or
   (2) Chang—post acquisition.
d. Software program for Quantification available through MicroDELTA User's Group.

SPECTRUM Medical Systems

Patient Preparation:

a. Subdued light; quiet area; and eyes closed.
b. Inject: 3mCi IMP (P,5n)
   1. wait 20 minutes before scanning
   2. 4 hours for delayed image

Acquisition:

a. Collimator: LEGP (1.3 cm system resolution 4.8 Kcps/μCi/ml or 570 cpm/μCi)
b. Matrix: 64 x 64
c. Sampling: 64 steps (5.625°)
d. Time/Step: 25 seconds (4 detector head aquisition)
e. Cts/Pixel Max: Early 150; Delayed 110
f. Orbit Time: 26 minutes
g. Magnification: x1 (17 x 22 cm UFOV)

Processing:

a. Pre-Data Filter: Wiener
b. Post Image Filter: None used
c. Display Format: 1024 x 1024
d. Attenuation Correction: Changes post correction
e. T ½ correction: Not used

TOSHIBA Medical Systems

Patient Preparation:

a. Patient's head should be stabilized using the velcro straps provided with the standard head holder.

b. Patient should be centered under the detector using the positioning lights built into the gantry.
c. Height of the ECT table should be adjusted so that the projection light intersects the midline of the patient's head.
d. Velcro strap should be placed across the patient's chin to tilt the head downward.

Acquisition:

a. Acquire study in 64^2 matrix for 360 degrees starting in the posterior projection.
b. Use Low Energy General Purpose collimator with projection images collected at every 6° angle interval for 40 seconds per sampling acquisition.
c. Average radius of rotation should be 11 to 13 centimeters.

Processing:

a. Projection data should be corrected for camera nonuniformities and smoothed prior to reconstruction.
b. After applying uniformity correction and a 5 x 5 pre-smooth, reconstruction should be performed through a combination of the convolution technique and the filtered backprojection technique utilizing the Shepp and Logan filter.
c. Attenuation correction should then be applied to thee reconstructed transaxial images by placing a movable ROI around a select cut.
d. A 3 x 3 smooth should be applied during sagittal and coronal tomographic slice generation.
e. Slice thickness may vary from 1-3 pixels, depending on preference (A slice thickness of 2 pixels is suggested).

SECTION III:

Bibliography
Index

Bibliography

A Ackerman RH: Of cerebral blood flow and SPECT (editorial). *Stroke* 15: 1-4; 1984.

Ackerman RH, Alpert NM, Correia JA, et al: Positron emission tomography in cerebrovascular ischemic disease. In *Functional Radionuclide Imaging of the Brain.* Magistretti PL, ed. New York, Raven Press: 277-280; 1983.

B Baldwin RM, Lin TH, Wu JL, et al: Pharmacology of amphetamines. In *Amphetamines and pH-Shift Agents for Brain Imaging.* Biersack HJ, Winkler C, eds. Berlin, Walter de Gruyter and Company: 3-10; 1986.

Baldwin RM, Lin TH, Wu JL, et al: Receptors for amphetamines. In *Amphetamines and pH-Shift Agents for Brain Imaging.* Biersack HJ, Winkler C, eds. Berlin, Walter de Gruyter and Company: 19-23; 1986.

Baldwin RM, Wu JL, Lin TH, et al: Pharmacokinetics of I-123 N-iosoproyl-p-iodo amphetamine (IMP) in humans. In *Current Applications in Radiopharmacology.* Billinghurst MW, ed. Toronto, Pergamon Press: 35-41; 1986.

Baldwin RM, Wu JL: In vivo chemistry of iofetamine HCL iodine—123 (IMP). *J Nucl Med* 29:122-124; 1988.

Biersack HJ, Fröscher W, Klünenberg H, et al: Brain SPECT with I-123-isopropyl amphetamine in epilepsy. *NucCompact* 14: 62-72; 1983.

Biersack HJ, Fröscher W, Penin H, et al: Brain SPECT with ^{123}I-labeled amphetamine derivatives in epilepsy and migraine. In *Amphetamines and ph-Shift Agents for Brain Imaging.* Biersack HJ, Winkler C, eds. Berlin, Walter de Gruyter and Company: 149-155; 1986.

Biersack HJ, Harmann A, Friedrich G, et al: Concerning the cause of crossed cerebellar diaschisis in cerebral vascular disease. *Nuklearmedizin* 23: 227-230; 1984.

Biersack HJ, Klünenberg H, Friedrich G, et al: ^{123}I-N-(sec. Butyl)-p-iodoamphetamine—a new radiopharmaceutical for brain imaging. *NucCompact* 15: 124-128; 1984.

Biersack HJ, Kreiten K, Harmann A, et al: ^{123}I-amphetamine SPECT in the diagnosis of neurological disorders. *Der Nuklearmediziner* 8: 57-63: 1985.

Biersack HJ, Penin H, Harmann A, et al: Cerebral SPECT with ^{123}I-labeled amphetamine. *Dtsch Med Wschr* 109: 1155-1159; 1984.

Bischof-Delaloye A, Delaloye B: Biokinetics of N-isopropyl, p-^{123}I-iodoamphetamine in the human. In *Amphetamines and pH-Shift Agents for Brain Imaging.* Biersack HJ, Winkler C, eds. Berlin, Walter de Gruyter and Company: 45-50; 1986.

Blau M: Radiotracers for functional brain imaging. *Semin Nucl Med* 15: 329-334; 1985.

Boyko OB, Park HM, Edwards MK, et al: I-123 HIPDM SPECT imaging and cerebral angiography for EC-IC bypass evaluation. *Radiographics* 5: 563-577; 1987.

Brott TG, Gelfand MJ, Williams CC, et al: Frequency and patterns of abnormality detected by iodine-123 amine emission CT after cerebral infarction. *Radiology* 158: 729-734; 1986.

Buell U, Krappel W, Schmiedek P, et al: Regional iodoamphetamine (^{123}I-IMP) uptake (SPECT) and regional cerebral blood flow (^{133}Xe-DSPECT). In *Amphetamines and pH-Shift Agents for Brain Imaging.* Biersack HJ, Winkler C, eds. Berlin, Walter de Gruyter and Company: 127-137; 1986.

Buell U, Olteanu-Nerbe V, Kirsch CM, et al: ^{123}I-amphetamine single photon emission computed tomography (SPECT) versus ^{133}Xe dynamic SPECT: preliminary results. *Nucl Med Comm* 5: 439-449; 1984.

Burt RW, Reddy RV, Mock BM, et al: Acetazolamide enhancement of HIPDM brain flow distribution imaging. *J Nucl Med* 27: 1627-1631; 1986.

Bushnell DL, Gupta S, Mlcoch AG, et al: Demonstration of focal hyperemia in acute cerebral infarction with iodine-123 iodoamphetamine. *J Nucl Med* 28:1920-1923; 1987.

C

Carlsen L, Andresen K: ^{131}I-labeled N-isopropyl-p-iodoamphetamine. *Eur J Nucl Med* 7: 280-281; 1982.

Carr EA: Scintigraphy and nuclear magnetic resonance: their potential in pharmacology; new radiopharmaceuticals and spin drugs: their potential implications for scintigraphy and nuclear magnetic resonance. *Clin Pharmacol Ther* 35: 94-108; 1984.

Cesaro P, Moretti JL, Caron JP, et al: Cerebral radionuclide tomography using ^{123}I-isopropyl-amphetamine in cerebral ischemic pathology. *Presse Med* 14: 205-208; 1985.

Cohen MB, Graham LS, Lake R, et al: Diagnosis of Alzheimer's disease and multiple infarct dementia by tomographic imaging of iodine-123 IMP. *J Nucl Med* 27: 769-774; 1986.

Coleman RE, Drayer BP, Jaszczak RJ: Studying regional brain function: A challenge for SPECT. *J Nucl Med* 23: 266-270; 1982.

Coni NK, Wraight EP, Barber RW: Regional cerebral perfusion in the elderly. *Age and Aging* 13: 214-217; 1984.

Creutizig H, Schober O, Gielow P, et al: Cerebral dynamics of N-isopropyl-(^{123})p-iodoamphetamine. *J Nucl Med* 27: 178-183; 1986.

D

deBruine JF, Van Royen EA, Van Weeren F, et al: Functional brain imaging with I 123-amphetamine, first experience in the Netherlands. *Clin Neurol Neurosurg* 88: 253-261; 1986.

deBruine JF, Van Royen EA, Vyth A, et al: Thallium-201 diethyldithiocarbamate: An alternative to iodine-123 N-isopropyl-p-iodoamphetamine. *J Nucl Med* 26: 925-930; 1985.

Defer G, Moretti J, Cesaro P, et al: Early and delayed SPECT using N-isopropyl p-iodoamphetamine iodine-123 in cerebral ischemia. *Arch Neurol* 44: 715-718; 1987.

DeLand FH: Perspectives. In *Functional Radionuclide Imaging of the Brain.* Magistretti PL, ed. New York, Raven Press: 347-356; 1983.

Devous Sr MD. Stokely EM, Bonte FJ: Quantitative imaging of regional cerebral blood flow by dynamic single photon tomography. In *Radionuclide Imaging of the Brain.* Holman BL, ed. New York, Churchill Livingston: 135-161; 1985.

E

Eckelmann WC, Reba RC, Rzeszotarski WJ, et al: External imaging of cerebral muscarinic acetlycholine receptors. *Science* 223: 291-293; 1984.

Ell PJ, Cullum I, Donaghy M, et al: Cerebral blood flow studies with ^{123}Iodine-labeled amines. *Lancet:* 1348-1352; 1983.

Ell PJ, Hocknell JML, Jarritt PH, et al: A ^{99}Tc^{m}-labeled radiotracer for the investigation of cerebral vascular disease. *Nucl Med Comm* 6:437-441; 1985.

Ell PJ, Jarritt PH, Costa DC, et al: Functional imaging of the brain. *Semin Nucl Med* 17: 214-229; 1987.

Ell PJ, Costa DC, Harrison M: Imaging cerebral damage in HIV infection. *Lancet* I: 569-570; 1987.

Engel Jr J, Troupin AS, Crandall PH, et al: Recent developments in the diagnosis and therapy of epilepsy. *Ann Intern Med* 97: 584-598; 1982.
Esser PD: Improvements in SPECT technology for cerebral imaging. *Semin Nucl Med* 15: 335-346; 1985.
Etani H, Kimura K, Yoneda S, et al: A study of cerebral blood flow dynamics in internal cartoid occlusion with N-isopropyl-^{123}I-p-iodoamphetamine (^{123}I-IMP). *No Shinkei* 37: 965-972; 1985.

F Falls M, Park CH, Madsen M: Iofetamine HCl I-123 (Iodoamphetamine) brain SPECT atlas. *Clin Nucl Med* 10: 443-449; 1985.
Fill H, Rössler H, Riccabona G, et al: ^{123}I-amphetamine (IMP) SPECT in patients with schizophrenia. *Nuklearmedizin* 22: 481-485; 1984.
Fuller RW: Structure-activity relationships among the halogenated amphetamines. *Ann NY Acad Sci* 305: 147-159; 1978.

H Hedde JP, Claussen C, Fiegler W, et al: Measurement of regional cerebral blood flow as a supplement to carotid angiography in the evaluation of cerebrovascular insufficiency. *Fortschr Rontgenstr* 142: 159-166; 1985.
Hedde JP, Fiegler W, Schulz S, et al: Comparison between ^{133}Xenon and ^{123}I-isopropyl amphetamine in the measurement of regional cerebral blood flow. *Fortschr Rontgenstr* 141: 212-219; 1984.
Henriksen L, Vorstrup S, Paulson OB: Regional cerebral blood flow measured by Xenon-133 and [^{123}I] iodoamphetamine in patients with cerebrovascular diseases. *Prog Brain Res* 62: 245-252; 1984.
Hellman RS, Collier BD, Tikofsky RS, et al: Comparison of single photon emission computed tomography with [^{123}I] iodoamphetamine and Xenon enhanced computed tomography for assessing regional cerebral blood flow. *J Cereb Blood Flow Metab* 6:747-755; 1986.
Hellman RS, Collier BD: MRI and SPECT I-123 iodoamphetamine imaging of a patient with progressive dementia. *Advances in Functional Neuroimaging* 1:18-20, 1988.
Hellman RS, Collier BD: Single photon emission computed tomography: A clinical experience. In *Nuclear Medicine Annual 1987.* Freeman LM, Weissman HS, eds. New York, Raven Press: 51-101-; 1987.
Hellman RS, Tikofsky RS, Collier BD, Joestgen TM, eds: A new era in functional brain imaging in the evaluation of stroke: A monograph. New York, Medi-Physics: 1988.
Higa T, Ito H, Saiki Y, et al: Cerebral blood flow imaging using N-isopropyl-p-I-123-iodoamphetamine (IMP). *Kaku Igaku* 22: 415-421; 1985.
Higa T, Tanaka T, Ikekubo K, et al: SPECT with N-isopropyl-p-iodoamphetamine in occlusive cerebrovascular diseases. *Clin Nucl Med* 11: 855-859; 1986.
Hill TC, Holman BL: Cerebral perfusion imaging. In *Nuclear Medicine Yearbook 1984.* Hoffer PB, ed. Chicago, Yearbook Medical Publishers, Inc.: 13-23; 1984.
Hill TC, Holman BL: SPECT brain imaging: finding a niche in neurologic diagnosis. *Diagnostic Imaging* 7: 64-68; 1985.
Hill TC, Holman BL, Lovett R, et al: Initial experience with SPECT (single photon emission computerized tomography) of the brain using N-isopropyl I-123 p-iodoamphetamine: Concise communication. *J Nucl Med 23:* 191-195; 1982.
Hill TC, Holman BL, Magistretti PL: Emission computed tomography of the brain with radiolabeled amines. In *Computed Emission Tomography.* Ell PF, Holman BL, eds. Oxford, Oxford University Press: 419-437; 1983.

Hill TC, Magistretti PL, Holman BL, et al: Assessment of regional cerebral blood flow (rCBF) in stroke using SPECT and N-isopropyl-(I-123)-p-iodoamphetamine (IMP). *Stroke* 15: 40-45; 1984.

Holman BL: Anatomy and function of the brain. In *Radionuclide Imaging of the Brain.* Holman BL, ed. New York, Churchill Livingston: 1-26; 1985.

Holman, BL: Perfusion and receptor SPECT in the dementias: George Taplin memorial lecture. *J Nucl Med* 27: 855-860; 1986.

Holman BL, Gibson RE, Hill TC, et al: Muscarinic acetylcholine receptors in Alzheimer's disease, in vivo imaging with iodine 123-labeled 3-quinuclidinyl-4-iodobenzilate and emission tomography. *JAMA* 254: 3063-3066; 1985.

Holman BL, Hill TC, Lee RGL, et al: Brain imaging with radiolabeled amines. In *Nuclear Medicine Annual 1983,* Freeman LM, Weissmann HS, eds. New York, Raven Press: 131-165; 1983.

Holman BL, Hill TC, Magistretti PL: Brain imaging emission computed tomography and radiolabeled amines. *Invest Radiol* 17: 206-215; 1982.

Holman BL, Hill TC, Polak JF, et al: Cerebral perfusion imaging with iodine 123-labeled amines. *Arch Neurol* 41: 1060-1063; 1984.

Holman BL, Hill TC, Polak JF, et al: Functional brain imaging with I-123-labeled amines. In *Radionuclide Imaging of the Brain.* Holman BL, ed. New York, Churchill Livingston: 163-184; 1985.

Holman BL, Jolesz FA, Polak JF, et al: Comparison of I-123 IMP cerebral uptake and MR spectroscopy following experimental carotid occlusion. *Invest Radiol* 20: 370-373; 1985.

Holman BL, Lee RG, Hill TC, et al: A comparison of two cerebral perfusion tracers, N-isopropyl I-123 p-iodoamphetamine and I-123 HIPDM, in the human. *J Nucl Med* 25: 25-30; 1984.

Holman BL, Magistretti PL, Hill TC, et al: Regional cerebral blood flow measured with N-isopropyl-123-I-p-iodoamphetamine. In *Functional Radionuclide Imaging of the Brain,* Magistretti PL, ed. New York, Raven Press: 217-224; 1983.

Holman BL, Wick MM, Kaplan M, et al: The relationship of eye uptake of N-isopropyl-p[^{123}I] iodoamphetamine to melanin production. *J Nucl Med* 25: 315-319; 1984.

Holman BL, Zimmerman RE, Schapiro JR, et al: Biodistribution and dosimetry of N-isopropyl-p-[^{123}I] iodoamphetamine in the primate. *J Nucl Med* 24: 922-931; 1983.

I Itoh K, Takei H, Tsukamoto E, et al: Clinical findings on N-isopropyl-p-[^{123}I] iodoamphetamine in cerebrovascular disorders. *Kaku Igaku* 22: 1727-1734; 1985.

J Jagust WJ, Budinger TF, Reed BR. The diagnosis of dementia with single photon emission tomography. *Arch Neurol* 44: 258-262; 1987.

Johnson KA, Mueller SP, Washe TM, et al: Cerebral perfusion imaging in Alzheimer's disease: Use of single photon emission computed tomography and iofetamine hydrochloride I-123. *Arch Neurol* 44: 165-168; 1987.

Johnson KA, Holman BL, Mueller SP, et al: Single photon emission computed tomography in Alzheimer's disease: Abnormal iofetamine I-123 uptake reflects dementia severity. *Arch Neurol* 45: 392-396; 1988.

BIBLIOGRAPHY

K King MA, Schwinger RB, Penney BC, et al: Digital restoration of indium-111 and iodine-123 SPECT images with optimized Metz filters. *J Nucl Med* 27:1327-1336; 1986.

Kizuka H, Elmaleh DR, Brownell GL, et al: Synthesis and evaluation of p-iodophenterine (IP) as a brain perfusion agent. *Nucl Med Comm* 6: 49-56; 1985.

Knapp FF Jr, Srivastava PC: Potential new approaches for the development of brain imaging agents for single-photon applications. In *Amphetamines and pH-Shift Agents for Brain Imaging.* Biersack HJ, Winkler C, eds. Berlin, Walter de Gruyter and Company: 71-83; 1986.

Knapp WH, von Kummer R, Kubler W: Imaging of cerebral blood flow-to-volume distribution using SPECT. *J Nucl Med* 27: 465-470; 1986.

Knust EJ, Machulla HJ, Baldwin RM, et al: Synthesis of and animal experiments with, N-isopropyl-p-^{123}I-iodoamphetamine (IMP) and ^{18}F-3-deoxy-3-flouro-d-glucose (3-FDG) as tracers in brain and heart diagnostic studies. *Nuklearmedizin* 23: 31-34; 1984.

Kobayashi H, Hayashi M, Kawano H, et al: Cerebral blood flow studies using N-isopropyl I-123 p-iodoamphetamine. *Stroke* 16: 293-296; 1985.

Kuhl DE: Imaging local brain function with emission computed tomography. *Radiology* 150: 625-631, 1984.

Kuhl DE: Mapping local cerebral glucose metabolism in stroke, degenerative diseases, and epilepsy. In *Functional Radionuclide Imaging of the Brain.* Magistretti PL, ed. New York, Raven Press: 257-267; 1983.

Kuhl DE, Barrio JR, Huang SC, et al: Quantifying local cerebral blood flow by N-isopropyl-p-[^{123}I] iodoamphetamine (IMP) tomography. *J Nucl Med* 23: 196-203; 1982.

Kung HF: Brain radiopharmaceuticals. In *Radiopharmaceuticals: Progress and Clinical Perspectives,* Volume 1. Fritzberg AR, ed. Boca Raton, CRC Press, Inc.: 21-39; 1986.

Kung HF, Tramposch KM, Blau M: A new brain perfusion imaging agent: [I-123] HIPDM: N, N, N′-trimethyl-N′ [2-hydroxy-3-methyl-5-iodobenzyl]-1, 3-propanediamine. *J Nucl Med* 24: 66-72; 1983.

Kurosaki A, Murata H, Seki Y, et al: The usefulness of N-isopropyl-I-123-p-iodoamphetamine in detecting abnormalities of regional cerebral blood flow. *Kaku Igaku* 22: 1505-1513; 1985.

L La France ND, Wagner Jr, HN, Whitehouse P, et al: Decreased accumulation of isopropyl-iodoamphetamine [I-123] in brain tumors. *J Nucl Med* 22: 1081-1083; 1981.

Lamb JF, Baldwin RM, Lin TH: A commercial viewpoint for the production of new radiopharmaceuticals. In *Applications of Nuclear and Radiochemistry.* Lambrecht RM, ed. New York, Pergamon Press: 89-101; 1982.

Larsson SA, Bergstrand G, Berstedt H, et al: A special cut-off gamma camera for high resolution SPECT of the head. *J Nucl Med* 25: 1023-1030; 1984.

Lassen NA, Henriksen L, Holm S, et al: Cerebral blood flow tomography: Xenon-133 compared with isopropyl-amphetamine-iodine-123: Concise Communication. *J Nucl Med* 24: 17-21; 1983.

Lassen NA, Henriksen L, Holm S, et al: Cerebral blood flow tomography by SPECT (single emission computed tomography): Xenon-133 compared to isopropyl-amphetamine-iodine-123. *Ann Radiol* 26: 53-58; 1983.

Lear JL, Ackermann RF, Kameyama M, et al: Evaluation of [^{123}I] isopropyl-iodoamphetamine as a tracer for local cerebral blood flow using direct autoradiographic comparison. *J Cereb Blood Flow Metab* 2: 179-185; 1982.

Lee RGL, Hill TC, Holman BL, et al: Comparison of N-isopropyl I-123 p-iodoamphetamine brain scans using Anger camera scintigraphy and single-photon emission tomography. *Radiology* 145: 789-793; 1982.

Lee RGL, Hill TC, Holman BL, et al: N-isopropyl (I-123) p-iodoamphetamine brain scans with single-photon emission tomography: Discordance with transmission computed tomography. *Radiology* 145: 795-799; 1982.

Lee RGL, Hill TC, Holman BL, et al: Predictive value of perfusion defect size using N-isopropyl-(I-123)-p-iodoamphetamine emission tomography in acute stroke. *J Neurosurg* 61: 449-452; 1984.

Liewendahl K, Kairento AL, Pyrhönen S, et al: Localization of melanoma with radiolabeled monoclonal antibody fragments and iodoamphetamine. *Eur J Nucl Med* 12: 359-362; 1986.

Lister-James J: Single photon radiopharmaceuticals. In *Radionuclide Imaging of the Brain.* Holman BL, ed. New York, Churchill Livingstone: 75-99; 1985.

Lyall D: Method for the preparation of N-isopropyl-p-(^{123}I)-iodoamphetamine with an assessment of radiochemical purity. *Nucl Med Comm* 6: 485-488, 1985

M Machulla HJ, Knust EJ: Recent developments in the field of ^{123}I-radiopharmaceuticals. *Nuklearmedizin* 23: 111-118; 1984.

Madsen MT, Park CH: Enhancement of SPECT images by Fourier filtering the projection image set. *J Nucl Med* 26: 395-402; 1985.

Magistretti PL, Uren RF, Blume H, et al: Delineation of epileptic focus by single photon emission tomography. *Eur J Nucl Med* 7: 484-485; 1982.

Magistretti PL, Uren RF, Parker AJ, et al: Monitoring of regional cerebral blood flow by single photon emission tomography of I-123-N-isopropyl-iodoamphetamine in epileptics. *Ann Radiol* 26: 68-71; 1983.

Magistretti PL, Uren RF, Royal HD, et al: N-isopropyl-123-I-p-iodoamphetamine imaging in epilepsy. In *Functional Radionuclide Imaging of the Brain.* Magistretti PL, ed. New York, Raven Press: 247-251; 1983.

Matsuda H, Seki H, Ishida H, et al: Regional cerebral blood flow measurement by N-isopropyl-p-(^{123}I) iodoamphetamine. *Kaku Igaku [Nuclear Medicine]* 21: 445-448; 1984.

Matsuda H, Seki H, Sumiya H, et al: Quantitative cerebral blood flow measurements using N-isopropyl-(Iodine 123) p-iodoamphetamine and single photon emission computed tomography with rotating gamma camera. *Am J Phys Imag* 1: 186-194; 1986.

Mertens J, Vanryckeghem W, Bossuyt A, et al: Preparative HPLC purification, a key to pure ^{123}IMP showing high brain uptake. *Nucl Med Comm* 5: 705-710;1984.

Moore SC: Quantitative capabilities of single-photon emission computerized tomography. In *Functional Radionuclide Imaging of the Brain.* Magistretti PL, ed. New York, Raven Press: 177-192; 1983.

Moretti JL: The diagnosis and prognosis of cerebral vascular diseases using early and late SPECT with N-isopropyl-p-(I-123) amphetamine (IMP) on 35 patients. *Jpn J Nucl Med* 23: 1121-1124; 1986.

Moretti JL, Askienazy S, Raynaud C, et al: N-isopropyl-123-I-p-iodoamphetamine: An agent for brain imaging with single-photon emission computerized tomography. In *Functional Radionuclide Imaging of the Brain.* Magistretti PL, ed. New York, Raven Press: 231-246; 1983.

BIBLIOGRAPHY

Moretti JL, Holman BL, Delmon L, et al: Effect of antidepressant and narcoleptic drugs on N-isopropyl p-iodoamphetamine biodistribution in animals. *J Nucl Med* 28: 354-359; 1987.

Mueller SP, Polak JF, Kijewski MF, et al: Collimator selection for SPECT brain imaging: The advantage of high resolution. *J Nucl Med* 27: 1729-1738; 1986.

N Neary D, Snowden JS, Shields, RA, et al: Single photon emission tomography using ^{99m}Tc-Hm-PAO in the investigation of dementia. *J Neurol Neurosurg Psychiatry* 50: 1101-1109; 1987.

O O'Leary DH, Hill TC, Lee RG, et al: The use of ^{123}I-iodoamphetamine and single photon emission computed tomography to assess local cerebral blood flow. *AJNR* 4: 547-549; 1983.

Oyamada H. Terui S, Fukukita H, et al: Clinical evaluation of single photon emission computed tomography of the brain. *Eur J Nucl Med* 7: 439-443; 1982.

P Palmer DW, Rao SA: A simple method to quantitate iodine-124 contamination in iodine-123 radiopharmaceuticals. *J Nucl Med* 26: 936-940; 1985.

Podreka I, Holl K, Dal Bianco P, et al: Clinical relevance of N-isopropyl-(^{123}I) p-iodoamphetamine (IMP) SPECT brain imaging. In *Amphetamines and pH Shift Agents for Brain Imaging.* Biersack HJ, Winkler C, eds. Berlin Walter de Gruyter and Company: 109-125; 1986.

Podreka I, Suess E, Goldenberg G, et al: Limited experience with technetium 99m HM-PAO brain SPECT. *J Nucl Med* 28: 1657-1666; 1987.

Polak JF, English RJ, Holman BL: Performance of collimators used for tomographic imaging of I-123 contaminated with I-124. *J Nucl Med* 24: 1065-1069; 1983.

Polak JF, Mueller SP, Holman BL: Caudate infarction demonstrated by N-isopropyl-p-iodoamphetamine SPECT imaging using a rotating gamma camera. *Clin Nucl Med* 11: 718-720; 1986.

R Rapin JR, LePoncin-Lafitte M, Duterte D, et al: Iodoamphetamine as a new tracer for local cerebral blood flow in the rat: Comparison with isopropyl-iodoamphetamine. *J Cereb Blood Flow and Metab* 4: 270-274; 1984.

Rapin JR, LePoncin-Lafitte M, Duterte D, et al: Radiopharmacologic studies of isopropyl-iodoamphetamine. In *Functional Radionuclide Imaging of the Brain.* Magistretti PL, ed. New York, Raven Press: 209-216; 1983.

Raynaud C, Rancurel G, Samson Y, et al: Pathophysiologic study of chronic infarcts with I-123 isopropyl-iodoamphetamine (IMP): The importance of periinfarct area. *Stroke* 18: 21-29; 1987.

Royal HD, Hill TC, Holman BL: Clinical brain imaging with isopropyl-iodoamphetamine and SPECT. *Semin Nucl Med* 15: 357-376; 1985.

S Sato M, Kohama A, Fukuda A, et al: Moyamoya-like diseases associated with ventricular hemorrhages: Report of three cases. *Neurosurgery* 17: 260-266; 1985.

Sharp P, Gemmell H, Cherryman G, et al: Application of iodine-123-labeled isopropyl amphetamine imaging to the study of dementia. *J Nucl Med* 27: 761-768; 1986.

Shiozaki H, Kishikawa T, Ikeda J, et al: Clinical significance of cerebral blood flow imaging using N-isopropyl-p-^{123} I-iodoamphetamine, with particular reference to cerebrovascular impairment. *Kaku Igaku* 22: 449-458; 1985.

Shiozaki H, Kishikawa T, Ikeda J, et al: Evaluation of anastomosis of the superficial temporal artery and the middle cerebral artery using N-isopropyl-p-^{123}I-iodoamphetamine. *Kaku Igaku* 22: 229-232; 1985.

Som P, Oster ZH, Yamamoto K, et al: Some factors affecting the cerebral and extracerebral accumulation of N-isopropyl-p-iodoamphetamine (IAMP). *Int J Nucl Med Biol* 12: 185-196; 1985.

Sone T, Fukunaga M, Otsuka N, et al: [Scintigraphy of cerebral blood flow with N-isopropyl-p-^{123}I-iodoamphetamine in patients with cerebrovascular impairment.] *Kaku Igaku* 22: 1447-1459; 1985.

T Tanada S: A case demonstrating the usefulness of brain SPECT with ^{123}I-IMP in cerebral infarction. *Rinsho Hoshasen [Jpn J Clin Radiology]* 29: 1141-1142; 1984.

Tikofsky RS, Harsch HH, Goldstein MD, et al: CT, magnetic resonance imaging and SPECT iodine-123-iodoamphetamine imaging of a patient with cognitive deterioration. *Clin Nucl Med* 12: 463-465; 1987.

Touya JJ, Rahimian J, Grubbs DE, et al: A noninvasive procedure for in vivo assay of a lung amine endothelial receptor. *J Nucl Med* 26: 1302-1307; 1985.

Tramposch KM, Kung HF, Blau M: Radioiodine-labeled amines as brain imaging agents. In *Applications of Nuclear and Radiochemistry.* Lambrecht RM, ed. Oxford, Pergamon Press: 204-213; 1982.

U Uren RF, Magistretti PL, Royal HD, et al: Single-photon emission computed tomography: A method of measuring cerebral blood flow in three dimensions (preliminary results of studies in patients with epilepsy and stroke). *Med J Aust* 1: 411-413; 1983.

V Van Heertum RL, O'Connell RA: The evaluation of psychiatric disease with IMP cerebral SPECT imaging. *Advances in Functional Neuroimaging* 1: 4-11; 1988.

Van Heertum RL, Yudd AP: Brain imaging of progressive dementia in a young male patient. *Advances in Functional Neuroimaging* 1: 17-19; 1988.

Van Royen EA, de Bruime JF, Hill TC, et al: Cerebral blood flow imaging with thallium-201-diethyldithiocarbomate SPECT. *J Nucl Med* 28: 178-183; 1987.

von Schulthess GK, Ketz E, Schubiger PA, et al: Regional quantitative non-invasive assessment of cerebrel perfusion and function with N-isopropyl-[^{123}I] p-iodoamphetamine. *J Nucl Med* 26: 9-16, 1985.

von Schulthess GK, Meili A, Weder B, et al: Regional quantification of SPECT studies of the brain with N-isopropyl-[^{123}I] p-iodoamphetamine (IMP). In *Amphetamines and pH-Shift Agents for Brain Imaging.* Biersack HJ, Winkler C, eds. Berlin, Walter de Gruyter and Company: 139-147; 1986.

W Wada M, Ichiya Y, Katsuragi M, et al: Scintigraphic visualization of human malignant melanoma with N-isopropyl-p-[I-123] iodoamphetamine. *Clin Nucl Med* 10: 415-417; 1985.

Wada M, Ichiya Y, Kuwabara Y, et al: Clinical evaluation of N-isoptopyl-p-^{123}I iodoamphetamine in the diagnosis of cerebrovascular disorders—a comparative study with x-ray CT and positron CT. *Kaku Igaku* 22: 1477-1483; 1985.

Wagner Jr HN: Nuclear medicine. *JAMA* 256: 2096-2097: 1986.

Winchell HS: Radiopharmaceuticals for the study of the brain. In *Functional Radionuclide Imaging of the Brain.* Magistretti PL, ed. New York, Raven Press: 225-230; 1983.

Winchell HS, Baldwin RM, Lin TH: Development of I-123-labeled amines for brain studies: Localization of I-123 iodophenylalkyl amines in rat brain. *J Nucl Med* 21: 940-946; 1980.

Winchell HS, Horst WD, Braun L, et al: N-isopropyl-[^{123}I] p-iodoamphetamine: Single-pass brain uptake and washout; binding to brain synaptosomes; and localization in dog and monkey brain. *J Nucl Med* 21: 947-952; 1980.

Woronowicz EM, Eisner RL, Gullberg DJ, et al: Factors affecting single photon emission computed tomography image quality and recommended QC procedures. *General Electric Medical Systems Operations.* Milwaukee, General Electric Company: 1982.

Wraight EP, Barber RW, Crossland P, et al: Tomographic images of cerebral blood flow using a slant hole collimator. *Nucl Med Comm* 4: 259-263; 1983.

Y

Yen CK, Pollycove M, Crass R, et al: Portasystemic shunt fraction quantification with colonic iodone-123 iodoamphetamine. *J Nucl Med* 27: 1321-1326, 1986.

Index

A

B

C

L

M

N

O

P

Q

R

S

T

U

V

W

NEW OFFER

SPECT BRAIN IMAGING WITH IOFETAMINE HCL I 123:
Differential Diagnosis of Dementia and Other Neuropsychiatric Disorders

Ronald L.Van Heertum, M.D. and
Ronald S. Tikofsky, Ph.D.

A 24 x 36 inch wall chart on cerebral SPECT imaging is now available from Trivirum Publishing Company.

The chart includes color illustrations of normal surface and ventricular anatomy with full identification.

Complete normal serial SPECT scans are displayed in all three orthogonal planes to assist in the comparative study of the disease states given.

Case studies of dementias, epilepsy and psychiatric disorders are given as a guide to the types of images that can be encountered in this imaging tecchnique.

To order fill out and return this form.

TO ORDER: A check or money order in U.S. CURRENCY MUST accompany each order to cover the postage and handling.

U.S. ORDERS:

First wall chart ordered = $10.00

Additional charts ordered at the same time are available at $4.00 each.

Send_____additional charts @ $4.00 = _______

New York State residents add state/local taxes_______

OUTSIDE CONTINENTAL U.S.:

First wall chart ordered = $15.00

Additional charts ordered at the same time are available at $5.00 each.

Send_____additional charts @ $5.00 = _______

TOTAL ENCLOSED = _______

PLEASE PRINT:

Name:_______________________________

Mailing Address:_______________________________

____

City:_______________State_______________Zip_______________

Return check and order form to: TRIVIRUM PUBLISHING CO.
211 East 43rd Street
New York, N.Y. 10017